For the Rising Math Olympians

Elementary School Math Contests

500+ Challenging Math Contest Problems and Detailed Step-By-Step Solutions

Steven Doan
Jesse Doan

ISBN-13: 978-1974443208
ISBN-10: 1974443205

For information or questions, please contact the authors, Steven Doan and Jesse Doan, at risingmatholympians@gmail.com.

Table of Contents

Table of Contents

Introduction

In sixth grade, I started participating in MATHCOUNTS®* and the American Mathematics Competitions, the premier middle school math contests. After competing in the 2012 MathCounts Nationals Competition in Orlando, Florida, I was inspired by meeting the 223 competitors who shared my passion for math. Before this competition, I felt I was the only person from Maui, Hawaii who was truly enthusiastic about solving rigorous math problems. Because of the huge impact that math competitions have made on my life, I wanted to extend the same opportunity to Maui County's elementary school students. The experience motivated me to found the Maui Math Challenge, an elementary school math contest aimed at teaching problem-solving skills and sparking interest in young minds to excel in math. The purpose of the Math Challenge is to create a nurturing environment where excellence in math is celebrated.

This book covers 12 contest rounds and over 500 problems from the four years of the competition. The contest problems tested are in Number Theory, Algebra, Counting & Probability, and Geometry. We have included over 500 detailed explanations and step-by-step solutions for each contest problem. In addition, there are formulas, strategies, and tips at the end of the book for reference.

It is highly recommended that the students working through this book are proficient in addition, subtraction, multiplication and division. Furthermore, knowledge in pre-algebra would be helpful.

We hope this book will serve as a valuable resource for young ambitious students who desire to excel in math competitions. This book may also serve as an accessible tool for math team coaches who wish to teach their students advanced problem-solving skills in preparation for elementary and middle school math competitions.

Jesse Doan
August 2017

*MATHCOUNTS® is a trademark registered by the MATHCOUNTS Foundation.

Acknowledgements

First of all, we would like to thank the Lord for all of His wisdoms, favors, and blessings. We would also like to thank our parents and our siblings for their continuous support and encouragement. Lastly, we would like to thank our math team coaches for inspiring us to help other young mathletes.

Maui Math Challenge

The Maui Math Challenge was a math competition for elementary school students in grades 3, 4, and 5 from public, private, and home-schools throughout Maui County, Hawaii. The competitions took place in 2013, 2014, 2015 and 2016. It was founded and organized in 2012 by Jesse Doan.

The Maui Math Challenge drew inspiration from middle school math competitions such as MATHCOUNTS and the American Mathematics Competitions.

OUR MISSION:

1. To increase interest in mathematics;
2. To develop problem solving skills and teach advanced math concepts to interested students;
3. To interest students in STEM and mathematics careers; and
4. To give recognition to outstanding math students.

Contest Rules:

1. Calculators are not permitted.
2. Notes and reference materials are not permitted.
3. Students should bring their own pencils and erasers. Other supplies, including scratch paper, will be provided.
4. Talking and signaling among team members are permitted only during the Team Round.
5. Communication between coaches and students is prohibited during any of the rounds.

The competition consisted of three rounds: the **Individual Round**, the **Team Round** and the **Countdown Round**.

Individual Round

In the Individual Round, contestants individually solve a written round consisting of 20 multiple choice questions and 10 short answer questions with a time limit of 40 minutes. Each multiple choice question is worth 1 point and each short answer question is worth 2 points. Calculators are not permitted. There is no penalty for guessing and there is no partial credit. The highest score possible is 40 points.

Team Round

In the Team Round, up to four contestants solve a written round consisting of 10 short answer questions with a time limit of 20 minutes. Each question is worth 2 points toward the Team Score. In this round, contestants may discuss the problems among team members. Calculators are not permitted. There is no penalty for guessing and there is no partial credit. The highest score possible is 20 points.

Countdown Round

The Countdown Round is a fast-paced head-to-head competition, and is the final round used in determining individual rankings. It is the only oral round of the competition. The Top 8 Individual Round Contestants are chosen to participate in this round.

In a round, two contestants compete face to face in the Countdown Round. A short-answer problem will be posted on a projector and the two contestants race to answer the problem in the given time of 1 minute. When a contestant solves the problem, the contestant will press the buzzer and give an answer. The first person to buzz in the correct answer will receive a point. If a person buzzes in a wrong answer, that person will not have a second chance to buzz in and the other person will have the remaining time to answer. If nobody correctly answers by the end of 1 minute, no point will be given.

After pressing the buzzer, the contestant will have 5 seconds to give an answer. Any answer given after the 5 seconds will be disqualified.

The Countdown Round is an elimination round.

The first four rounds are the **Quarterfinals** where four contestants are eliminated and four contestants proceed. The contestant that answers the most out of 3 questions correctly will advance.*

The next two rounds are the **Semifinals** where two of the four remaining contestants are eliminated and the other two proceed. The first contestant to answer 3 questions correctly will advance.

The second-to-last round is used to determine **3rd Place** in the Countdown Round. The two contestants that were eliminated in the Semifinals will participate in this round. The first contestant that answers 3 questions correctly wins 3rd Place.
The last round is the **Finals** and the two remaining contestants go head-to-head to determine **1st Place** and **2nd Place**. The first contestant that answers 4 questions correctly wins 1st Place and the other contestant wins 2nd Place.

*If there is a tie in the Quarterfinals, the contestant that advances is the one who is the first to answer a question correctly after the initial three questions.

2013
Maui Math
Challenge

[This page is intentionally left blank]

MAUI MATH CHALLENGE

2013
■ Individual Round ■
Problems 1-30

Name ____

School Name ______________________________ Grade __________

DO NOT BEGIN UNTIL YOU ARE INSTRUCTED TO DO SO.

1. The Individual Round consists of 20 multiple choice problems and 10 short answer problems.

2. You will have **40 minutes** to complete them.

3. Each Multiple-Choice Problem is worth 1 point and each Short Answer Problem is worth 2 points. There is no partial credit. The maximum score is 40 points.

Multiple Choice	Short Answer	Total Score	Scorer's Initials
+	×2	=	
+	×2	=	

2013 Individual Round Answer Sheet

1.	(a)	(b)	(c)	(d)	(e)	21. ________
2.	(a)	(b)	(c)	(d)	(e)	
3.	(a)	(b)	(c)	(d)	(e)	22. ________
4.	(a)	(b)	(c)	(d)	(e)	
5.	(a)	(b)	(c)	(d)	(e)	23. ________
6.	(a)	(b)	(c)	(d)	(e)	
7.	(a)	(b)	(c)	(d)	(e)	24. ________
8.	(a)	(b)	(c)	(d)	(e)	
9.	(a)	(b)	(c)	(d)	(e)	25. ________
10.	(a)	(b)	(c)	(d)	(e)	
11.	(a)	(b)	(c)	(d)	(e)	26. ________
12.	(a)	(b)	(c)	(d)	(e)	
13.	(a)	(b)	(c)	(d)	(e)	27. ________
14.	(a)	(b)	(c)	(d)	(e)	
15.	(a)	(b)	(c)	(d)	(e)	28. ________
16.	(a)	(b)	(c)	(d)	(e)	
17.	(a)	(b)	(c)	(d)	(e)	29. ________
18.	(a)	(b)	(c)	(d)	(e)	
19.	(a)	(b)	(c)	(d)	(e)	30. ________
20.	(a)	(b)	(c)	(d)	(e)	

2013 Individual Round Problems

1. What is the perimeter of a square with side length 6?

a) 24 b) 12 c) 36 d) 34 e) 30

2. Find the perfect cube of the first prime number.

a) 0 b) 1 c) 4 d) 8 e) 27

3. What is the positive difference between 2^3 and 3^2?

a) 2 b) 0 c) -1 d) 3 e) 1

4. Find the mean of the set: {22, 22, 28, 29, 34}.

a) 34 b) 22 c) 28 d) 27 e) 26

5. Using the approximation $\pi = \frac{22}{7}$, what is the area of a circle with radius 7?

a) 44 b) 14π c) 144 d) 154 e) 164

6. A regular die is rolled. What is the probability that a prime number is rolled?

a) ½ b) ⅓ c) ⅔ d) ⅚ e) 1

7. How many quarters are needed to pay for a $6.75 toll?

a) 23 b) 25 c) 27 d) 29 e) 31

8. Which of the following numbers is divisible by 8?

a) 25,412 b) 34,612 c) 78,512 d) 56,388
e) 49,122

9. Find the sum: $1 + 2 + 3 + \ldots + 9 + 10 + 11$.

a) 45 b) 55 c) 56 d) 65 e) 66

10. Joe has 35 minutes for his lunch break. He left his office from 10:17 AM to 11:08 AM. How many minutes did he go over his lunch break?

a) 26 minutes b) 91 minutes c) 51 minutes d) 16 minutes
e) 61 minutes

11. The ratio of boys to girls in a classroom is 7:5. If there are 60 girls in the classroom, how many boys are there?

a) 35 boys b) 64 boys c) 84 boys d) 25 boys
e) 42 boys

12. How many diagonals does a regular pentagon have?

a) 10 diagonals b) 5 diagonals c) 25 diagonals d) 6 diagonals
e) 7 diagonals

13. The President of a given planet gets paid $400,000 a year. If this planet has 20 months per year, 25 days per month and 32 hours per day, how much does the President get paid per hour?

a) $25 per hour b) $100 per hour c) $800 per hour d) $50 per hour
e) $200 per hour

14. The probability of having a boy or a girl is equally likely. A family has three children. What is the probability that the family has at least one boy?

a) $\frac{5}{8}$ b) $\frac{5}{7}$ c) $\frac{6}{7}$ d) $\frac{3}{4}$ e) $\frac{7}{8}$

15. Laura and Lucy live 540 miles apart. They both leave their house at the same time and drive toward each other. If Laura is driving at a rate of 45 miles per hour and Lucy is driving at a rate of 15 miles per hour, how many hours will it take for them to meet?

a) 12 hours b) 9 hours c) 36 hours d) 18 hours
e) 6 hours

16. Using only pennies and quarters, what is the least number of coins necessary to make 68 cents?

a) 7 coins b) 10 coins c) 19 coins d) 20 coins
e) 44 coins

17. What is the value of $32^2 - 28^2$?

a) 230 b) 240 c) 248 d) 250 e) 260

18. A right triangle has leg 8 and hypotenuse 17. What is the area of the triangle?

a) 40 b) 60 c) 68 d) 120 e) 136

19. A coin is flipped three times. What is the probability that exactly two of the coins are heads?

a) ⅛ b) ⅜ c) ¾ d) ⅞ e) ⅚

20. What is the Least Common Multiple (LCM) of 65 and 91?

a) 156 b) 455 c) 5915 d) 26 e) 910

21. John has three more than half of the number of pencils that James has. If James has 10 pencils, how many pencils does John have?

22. What is the product of the first 5 natural numbers?

23. Anna's outfit consists of one shirt and one pair of pants. If she has 3 different pairs of pants and 7 different shirts, how many combinations of outfits does she have?

24. If the first term of an arithmetic sequence is 0, the second term is 5 and the third term is 10, what is the product of the first term and the 2013th term?

25. Each school bus has 23 passenger seats. If 1,000 students go on a field trip and ride the school bus, what is the least number of buses needed?

26. How many prime numbers are between 1 and 50?

27. Let *n!* = n × (n – 1) × (n – 2) × … × 3 × 2 × 1. For example, *4!* = 4 × 3 × 2 × 1 = 24. What is the tens digit of *7!* ?

28. How many ways can the letters in the word "BOOT" be arranged?

29. There are ten people in a room. If each person shakes hands with everyone else once, how many handshakes are there? (Note: A person does not shake hands with him/herself)

30. The set {a, b, c} is given such that: a < b < c. The sums (a + b), (a + c), and (b + c) are equal to 10, 11, and 15, respectively. What is the value of (c – a)?

MAUI MATH CHALLENGE

2013
■ Team Round ■
Problems 1-10

Name ______________________ Grade __________

Name ______________________ Grade __________

Name ______________________ Grade __________

Name ______________________ Grade __________

School Name ______________________

DO NOT BEGIN UNTIL YOU ARE INSTRUCTED TO DO SO.

1. The Team Round consists of 10 short answer problems.

2. You will have **20 minutes** to complete them.

3. Each problem is worth 2 points toward your team score. There is no partial credit. The maximum score on the Team Round is 20 points.

Total Correct	Total Score	Scorer's Initials
$\times 2$	$=$	
$\times 2$	$=$	

2013 Team Round Answer Sheet

1. ______________________________

2. ______________________________

3. ______________________________

4. ______________________________

5. ______________________________

6. ______________________________

7. ______________________________

8. ______________________________

9. ______________________________

10. ______________________________

2013 Team Round Problems

1. What is the remainder when 5^6 is divided by 2?

2. All of the 17 letters in MAUI MATH CHALLENGE are placed in a pot. If one letter is selected at random, what is the probability that it is an A or an L? Express your answer as a common fraction.

3. The four side lengths of a square are $(6x - 2)$, $(3x + 10)$, $(4x + 6)$, and $(7x - 6)$. What is the perimeter of the square? Express your answer as a positive integer.

4. The edge lengths of a rectangular solid are 3, 5 and 11. What is the positive difference between the numerical value of the volume and the numerical value of the total surface area?

5. A car travels 2310 feet. It travels at a constant rate of 66 inches per second. How many minutes does it take the car to reach its destination?

6. A sandwich consists of one kind of bread, one kind of cheese, one kind of vegetable and one kind of meat. If there are three different breads, four different cheeses, two different vegetables, and three different meats, how many unique sandwiches can one make?

7. The number 1001 = 7 × 11 × 13. What is the greatest prime factor of 183,183?

8. Given the set: {20, 20, 21, 23, 25, 28, 31}, what is the sum of the set's mean, mode and range?

9. Find the value of $\dfrac{1+2+3+\ldots+98+99+100}{101}$.

10. How many ways can the letters in the word ARRAY be arranged?

MAUI MATH CHALLENGE

2013
■ Countdown Round ■
Problems 1-80

---

1. The Countdown Round consists of 80 short answer problems.

2. You will have **1 minute** to complete each problem.

3. Each problem is worth 1 point.

2013 Countdown Round Answer Sheet

1. ____________
2. ____________
3. ____________
4. ____________
5. ____________
6. ____________
7. ____________
8. ____________
9. ____________
10. ____________
11. ____________
12. ____________
13. ____________
14. ____________
15. ____________
16. ____________
17. ____________
18. ____________
19. ____________
20. ____________
21. ____________
22. ____________
23. ____________
24. ____________
25. ____________
26. ____________
27. ____________
28. ____________
29. ____________
30. ____________
31. ____________
32. ____________
33. ____________
34. ____________
35. ____________
36. ____________
37. ____________
38. ____________
39. ____________
40. ____________
41. ____________
42. ____________
43. ____________
44. ____________
45. ____________
46. ____________
47. ____________
48. ____________
49. ____________
50. ____________
51. ____________
52. ____________
53. ____________
54. ____________
55. ____________
56. ____________
57. ____________
58. ____________
59. ____________
60. ____________
61. ____________
62. ____________
63. ____________
64. ____________
65. ____________
66. ____________
67. ____________
68. ____________
69. ____________
70. ____________
71. ____________
72. ____________
73. ____________
74. ____________
75. ____________
76. ____________
77. ____________
78. ____________
79. ____________
80. ____________

2013 Countdown Round Problems

1. What is the largest two-digit number that is divisible by 7?

2. What percent of 10 is 5?

3. How many factors does the number 6 have?

4. What is the product of the first three prime numbers?

5. What is the area of a circle with radius 3? Express your answer in terms of π.

6. John had $5.00 when he entered a toy store. He spent 20% of his money on a toy truck and had exactly enough money to buy a toy car. How much did the toy car cost in dollars?

7. Find the value of $7^2 + 3 \times 17$.

8. Happy was so happy that he ate 25 jellybeans out of a jar. His sister, who was also happy, ate 50 jellybeans out of the jar. The rest, which is 25% of the original amount of jellybeans, was eaten by their mom. How many jellybeans were there in the jar originally?

9. A right triangle has legs 3 feet and 4 feet, and hypotenuse 5 feet. What is the area of the triangle in square feet?

10. If Mary can hold 6 cups in one hand, what percent of a box with 30 cups can she hold in two hands?

11. If one newspaper and two books cost \$7.00 while two newspapers and one book cost \$6.50, how much do one newspaper and one book cost in dollars?

12. How many seconds are there in ½ hour?

13. The cross country team ran 4.5 miles each day in February of 2013. How many miles did they run total in February 2013?

14. Find the value of $1 + 2 + 3 + \ldots + 8 + 9 + 10$.

15. Find the perimeter of a regular octagon with side length 8.

16. What is the average of the angles in a triangle with angles: 80°, 75° and 25°?

17. What is the product of the sum and the positive difference of the numbers 3 and 6?

18. If a tree is 2 yards, 4 feet and 5 inches tall, how tall is the tree in inches?

19. Find the positive difference of the two numbers 201.3 and 230.3.

20. What is the area of a square with perimeter 20?

2013 Countdown Round Problems

21. If Cindy starts eating lunch at 3:43 PM and she eats for 38 minutes, at what time does she finish eating?

22. Round π to the nearest hundredth.

23. Find the value of $4 \times 4 \times 4 + 6 \times 6$.

24. If Zach eats between 6 and 10 candies, how many different numbers of candies could he have eaten?

25. What is the Greatest Common Factor (GCF) of 12, 16 and 20?

26. Carlos wears one shirt and one pair of pants every day. If he has 8 shirts and 3 pairs of pants, how many different combinations of clothes can he wear?

27. What is the volume of a cube with side length 7 inches? Express your answer in cubic inches.

28. How many ounces are in 3.5 pounds?

29. What is the sum of the interior angles in a regular hexagon?

30. What is the remainder when 1,234,567,893 is divided by 5?

31. What is the Least Common Multiple (LCM) of 5 and 13?

32. How many yards are in a mile?

33. Find the volume of a rectangular solid with lengths of 6, 8 and 10.

34. How many days are in 312 hours?

35. When two coins are flipped, what is the probability that they are both heads? Express your answer as a common fraction.

36. A bag of gummy worms contains 3 yellow, 4 blue and 8 red worms. What is the probability of drawing a red worm? Express your answer as a common fraction.

37. An Algebra book costs $45 each. Each box contains 6 Algebra books. What is the total cost of the books if there are 5 boxes? Express your answer in dollars.

38. A bag has 3 red marbles, 6 blue marbles and 4 green marbles. What is the least number of marbles that James needs to take out in order to ensure that he has at least 1 red marble?

39. Buy-A-Cube-Box sells boxes that are cubes. If the side length of a box is 2 inches, what is the surface area of this closed box? Express your answer in square inches.

40. Solve for q: $7q - 31 = 18$.

41. What is the sum of all of the factors of 10?

42. Find the value of 17×23.

43. If $n = 4$, what is the value of $n^2 - 5n + 7$?

44. Joe takes 26 minutes to drive from home to work. If work starts at 11:37 AM, what time should Joe leave his house in order to arrive at work on time?

45. What is the circumference of a circle with area 9π? Express your answer in terms of π.

46. Find the value of $12{,}590 - 11{,}480 + 1$.

47. What is the range of the set: {31, 22, 58, 99, 79, 124}?

48. Solve for x: $13x - 3 = 36$.

49. It takes Celine one hour to travel 60 miles. How many minutes does it take Celine to travel 15 miles?

50. A suit sells for $100. If the manager decides to give a discount of 35% off on the suit, how much would the suit sell for now? Express your answer in dollars.

51. Find the area of a square with perimeter 16.

52. A movie starts at 11:47 PM and it runs for 203 minutes. At what time does the movie end?

53. What is six more than five times a dozen?

54. What is 15% of 60?

55. Jarvis can drive 5 miles in 20 minutes. How many hours does it take him to drive 45 miles?

56. Find the value of $[(2 \times 12) - 22]^2$.

57. What is the sum of the interior angles in a triangle?

58. If $A + B + C = 10$ and $A = C = 3$, what is the value of B?

59. What is the remainder when 2013×2015 is divided by 2?

60. If there are 15 buses and each bus can hold 23 students, what is the maximum number of students that the buses can hold?

61. What is the remainder when 7^2 is divided by 13?

62. Round $\frac{5}{33}$ to the nearest hundredth.

63. What is the area of a triangle with base 7 inches and height 18 inches? Express your answer in square inches.

64. How many yards are in 252 inches?

65. What is the mode in the set {11, 12, 14, 15, 18, 19, 15, 20}?

66. Robert goes to see a 169-minute movie. If the movie starts at 3:43 PM, what time does the movie end?

67. The letters in the words COUNTDOWN ROUND are placed in a pot. If a letter is chosen at random, what is the probability that a vowel is chosen? Express your answer as a common fraction.

68. What is half of 60% of 100?

69. What is the value of $2000^2 + 13^2$?

70. Johnny eats 1 apple on Monday. On Tuesday, he eats 2 apples and on Wednesday, he eats 3 apples. If he continues this pattern, how many apples will he have eaten in one week?

71. In the arithmetic sequence: {1, 4, 7, 10, … }, what is the value of the 7th term?

72. The probability that it rains on Monday is ⅕. The probability that it doesn't rain on Tuesday is $\frac{7}{12}$. What is the probability that it rains on both Monday and Tuesday? Express your answer as a common fraction.

73. How many minutes are in 6 days?

74. Solve for r. $7r - 2^4 = 75$.

75. What percent of 50 is 23?

76. What is the remainder when 13^{13} is divided by 13?

77. The area of a square is 36. If an equilateral triangle has the same perimeter as the square, what is the side length of the equilateral triangle?

78. Find the digit n that satisfies the following equation: $1{,}234{,}56\underline{n} = 123{,}456 \times 10$

79. What is the sum of the angles in a regular quadrilateral?

80. Find the value of $1 + 3 + 5 + \ldots + 11 + 13 + 15$.

2014 Maui Math Challenge

[This page is intentionally left blank]

MAUI MATH CHALLENGE

2014

■ Individual Round ■

Problems 1-30

Name ____

School Name ____________________________ Grade ________

DO NOT BEGIN UNTIL YOU ARE INSTRUCTED TO DO SO.

1. The Individual Round consists of 20 multiple choice problems and 10 short answer problems.

2. You will have **40 minutes** to complete them.

3. Each Multiple-Choice Problem is worth 1 point and each Short Answer Problem is worth 2 points. There is no partial credit. The maximum score is 40 points.

Multiple Choice	Short Answer	Total Score	Scorer's Initials
+	×2	=	
+	×2	=	

2014 Individual Round Answer Sheet

1. (a) (b) (c) (d) (e)
2. (a) (b) (c) (d) (e)
3. (a) (b) (c) (d) (e)
4. (a) (b) (c) (d) (e)
5. (a) (b) (c) (d) (e)
6. (a) (b) (c) (d) (e)
7. (a) (b) (c) (d) (e)
8. (a) (b) (c) (d) (e)
9. (a) (b) (c) (d) (e)
10. (a) (b) (c) (d) (e)
11. (a) (b) (c) (d) (e)
12. (a) (b) (c) (d) (e)
13. (a) (b) (c) (d) (e)
14. (a) (b) (c) (d) (e)
15. (a) (b) (c) (d) (e)
16. (a) (b) (c) (d) (e)
17. (a) (b) (c) (d) (e)
18. (a) (b) (c) (d) (e)
19. (a) (b) (c) (d) (e)
20. (a) (b) (c) (d) (e)

21. ____________________
22. ____________________
23. ____________________
24. ____________________
25. ____________________
26. ____________________
27. ____________________
28. ____________________
29. ____________________
30. ____________________

2014 Individual Round Problems

1. Solve: $6 + 2 \times (3^2 - 5)$.

a) 60 b) 34 c) 24 d) 14 e) 8

2. How many prime numbers are between 30 and 60?

a) 6 b) 7 c) 8 d) 9 e) 10

3. Given a set: {17, 33, 5, 25}. Find the median.

a) 17 b) 20 c) 21 d) 22 e) 25

4. What is the area of a square with perimeter 36 units?

a) 16 units^2 b) 324 units^2 c) 36 units^2 d) 64 units^2
e) 81 units^2

5. 485 pennies and 123 dimes have the same value as n nickels. What is the value of n?

a) 608 b) 343 c) 1715 d) 1100
e) 1443

6. Find the perimeter of the triangle shown below. Note: Figure not drawn to scale.

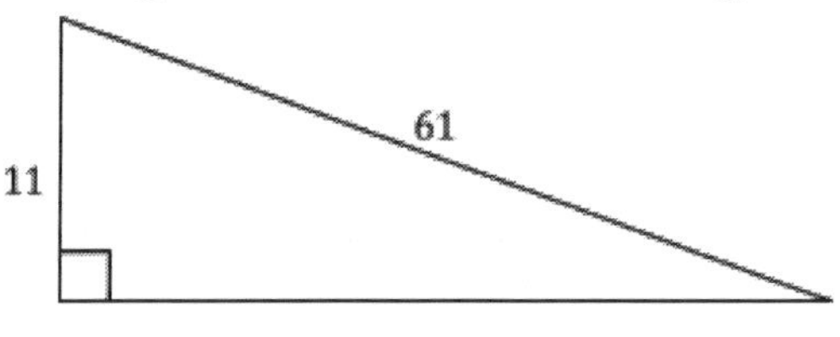

a) 132 b) 72 c) 83 d) 330 e) 660

7. Maui Math Cinemas is showing a movie that is 82 minutes in length. If the movie starts at 7:43 PM, what time will the movie finish?

a) 8:05 AM b) 9:05 AM c) 8:05 PM d) 9:15 PM
e) 9:05 PM

8. All the letters in the alphabet are placed in a bag and shuffled. If John reaches in the bag and draws one letter, what is the probability that he will draw a vowel? Assume that the letter "y" is not a vowel.

a) $\frac{5}{26}$ b) $\frac{21}{26}$ c) $\frac{1}{5}$ d) $\frac{21}{25}$ e) $\frac{3}{13}$

9. Josh has $10 more than a third of Jamie's money. If Jamie has $12, how much money does Josh have?

a) $22 b) $6 c) $14 d) $12 e) $46

10. Antonio takes a quick jog around a rectangular track with dimensions 300 feet by 400 feet. If he starts at one corner jogging at a speed of 50 feet per minute, how long does it take him to return back to his starting point?

a) 14 minutes b) 28 seconds c) 840 seconds d) 28 hours
e) 1680 seconds

11. Find the product of the Least Common Multiple (LCM) and the Greatest Common Factor (GCF) of 12 and 18.

a) 30 b) 6 c) 216 d) 36 e) 108

12. Solve. $\dfrac{5^3 - 7 * (3 + 8)}{10 - 2 * \sqrt{3} + 1}$

a) –8 b) 8 c) –16 d) 16 e) 0

13. Manny's daily outfit consists of one shirt and one pair of pants. If he has four different shirts and five different pairs of pants, how many unique outfits can he make?

a) 9 b) 10 c) 18 d) 19 e) 20

14. A regular die is rolled. What is the probability that the number rolled is a multiple of three?

a) ⅓ b) ⅔ c) ½ d) ⅙ e) ⅚

15. How many fluid ounces are in five gallons?

a) 160 fl. oz. b) 320 fl. oz. c) 640 fl. oz. d) 1280 fl. oz.
e) 2560 fl. oz.

16. A circular wheel has circumference 16π. Mary wants to buy four of these wheels for her car. What is the total area of all the wheels that she wants to purchase?

a) 16π b) 64π c) 128π d) 256π
e) 1024π

17. Which of the following numbers is divisible by 9?

a) 91,893 b) 78,238 c) 89,999 d) 75,735
e) 65,432

18. Find the area of the given rectangle shown below. Note: Figure not drawn to scale.

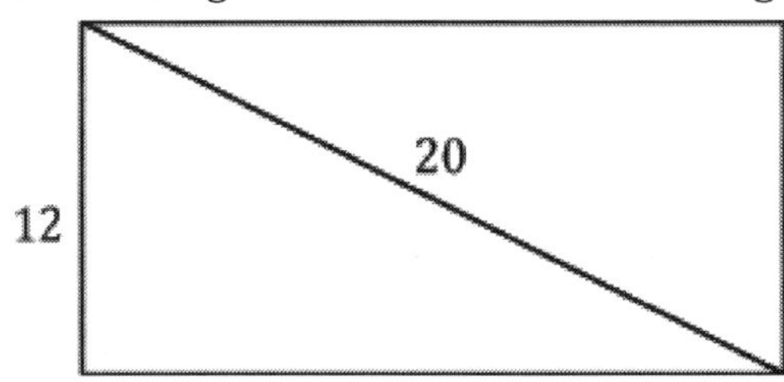

a) 240 b) 192 c) 56 d) 96 e) 384

19. Given a set: {23, 7, 29, 28, 13}. Find the positive difference between the mean and the median.

a) 2 b) 3 c) 9 d) 43 e) 49

20. The ratio of boys to girls at an elementary school is 4:5. If there are 45 girls in the school, how many boys are there?

a) 20 boys b) 25 boys c) 36 boys d) 45 boys
e) 60 boys

21. How many diagonals does a heptagon have?

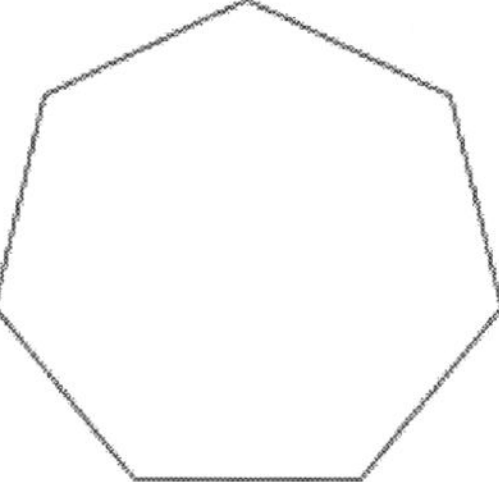

22. Samantha loves the number 5 so much that she wants to count how many times the digit 5 appears from 1 to 100. How many times does the digit 5 appear from 1 to 100?

23. Rebecca has to walk from home to school. The journey is a third of a mile. To keep herself busy, she calculates that she must travel n feet in order to go from home to school. What is the value of n?

24. Carl drives at a constant rate of 50 miles per hour from point A to point B. The distance he must travel is 125 miles. If he leaves point A at 12:00 PM, what time will he arrive at point B?

25. A classroom has 5 students: Al, Bob, Carrie, Don and Effie. How many different 2-person committees can be chosen from these 5 students? Note: A committee of Al-Bob is the same as a committee of Bob-Al.

26. Two fair coins are flipped. What is the probability that one head and one tail are flipped. Express your answer as a common fraction.

27. What is the surface area of a cube with volume 343 cubic meters? Express your answer in square meters.

28. Find the sum: $11 + 12 + 13 + \ldots + 39 + 40 + 41$.

29. What is the units digit of the decimal representation of 4^{11} ?

30. Pencils are sold in packs of twelve. If Sally needs to buy 1,000 pencils, what is the least number of packs of pencils that she must buy?

MAUI MATH CHALLENGE

2014
■ Team Round ■
Problems 1-10

Name ______________________________ Grade ____________

Name ______________________________ Grade ____________

Name ______________________________ Grade ____________

Name ______________________________ Grade ____________

School Name ______________________________

DO NOT BEGIN UNTIL YOU ARE INSTRUCTED TO DO SO.

1. The Team Round consists of 10 short answer problems.
2. You will have **20 minutes** to complete them.
3. Each problem is worth 2 points toward your team score. There is no partial credit. The maximum score on the Team Round is 20 points.

Total Correct	Total Score	Scorer's Initials
$\times 2$	$=$	
$\times 2$	$=$	

2014 Team Round Answer Sheet

1. ______________________________

6. ______________________________

2. ______________________________

7. ______________________________

3. ______________________________

8. ______________________________

4. ______________________________

9. ______________________________

5. ______________________________

10. ______________________________

2014 Team Round Problems

1. There are cows, chickens and pigs in a barn. The ratio of cows to chickens to pigs is 5:17:14. The number of chickens in the barn is 34. How many total animals are there in the barn?

2. Find the positive difference between the sum of the number of edges, the number of vertices and the number of faces of a cube, and the sum of the number of edges, the number of vertices and the number of faces of a tetrahedron.

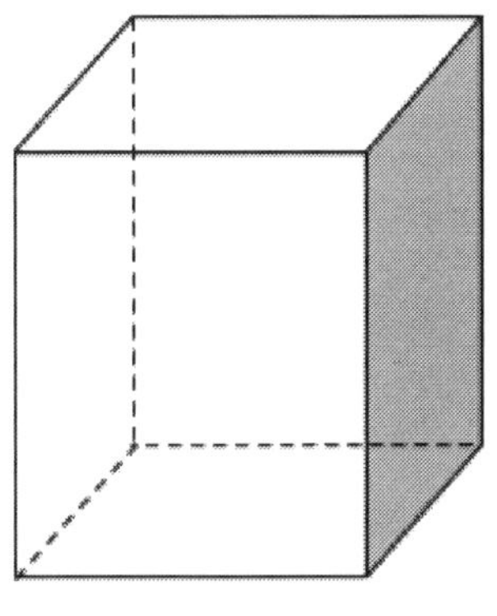

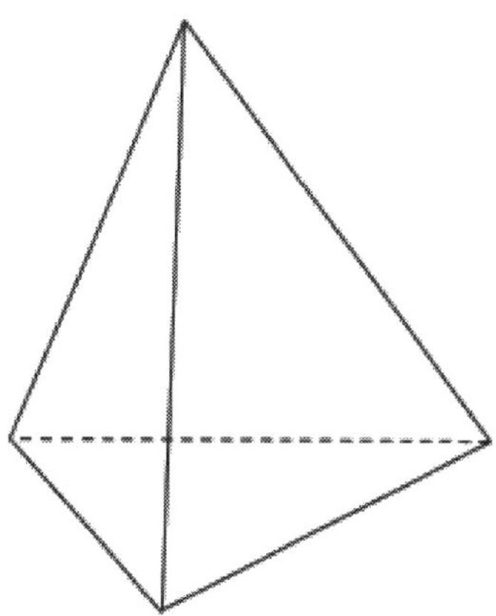

3. Find the product of the mean, median and range of the set: {2, 0, 2, 5, 5, 20, 11, 3, 3, 19, 7, 1, 13}.

4. How many ways can the letters in the word "PEOPLE" be arranged?

5. What is the remainder when 9^{12} is divided by 10?

6. A subway train in Washington D.C. departs every 2014 seconds. If a train just left at 9:55:36 AM, what time will the next subway train depart? Express your answer in the form AB:CD:EF where AB denotes the hour, CD denotes the minute and EF denotes the second.

7. Find the perimeter of the following figure if all of the angles are either 90 or 270 degrees. Note: Figure not drawn to scale.

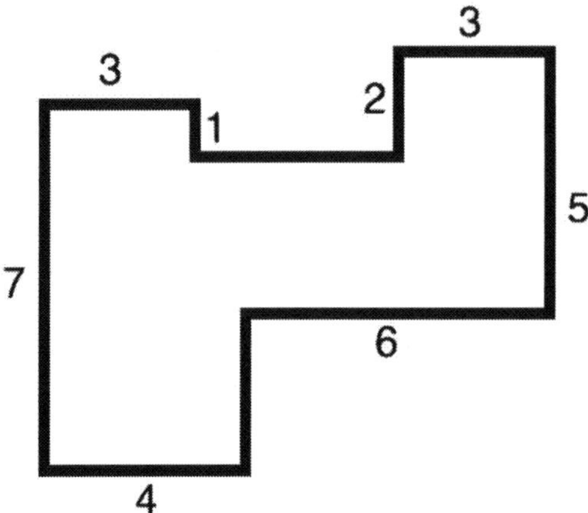

8. Assume that each letter in the alphabet is worth its place in the alphabetical order. For example, A = 1, B = 2, …, Y = 25 and Z = 26. If this is true, what is the sum of the values of the letters in the word ALPHABET?

9. Farmer Kai has only chickens and horses on his farm. In total, the animals have 18 heads and 62 legs. Assuming chickens have 2 legs and horses have 4 legs, how many chickens does Kai have?

10. Danny has 2 different shirts, 2 different pair of pants and 3 different hats. If an outfit consists of one shirt, one pair of pants and one hat, how many different outfits could Danny wear?

MAUI MATH CHALLENGE

2014
■ Countdown Round ■
Problems 1-80

---

1. The Countdown Round consists of 80 short answer problems.

2. You will have **1 minute** to complete each problem.

3. Each problem is worth 1 point.

2014 Countdown Round Answer Sheet

1. ________	21. ________	41. ________	61. ________
2. ________	22. ________	42. ________	62. ________
3. ________	23. ________	43. ________	63. ________
4. ________	24. ________	44. ________	64. ________
5. ________	25. ________	45. ________	65. ________
6. ________	26. ________	46. ________	66. ________
7. ________	27. ________	47. ________	67. ________
8. ________	28. ________	48. ________	68. ________
9. ________	29. ________	49. ________	69. ________
10. ________	30. ________	50. ________	70. ________
11. ________	31. ________	51. ________	71. ________
12. ________	32. ________	52. ________	72. ________
13. ________	33. ________	53. ________	73. ________
14. ________	34. ________	54. ________	74. ________
15. ________	35. ________	55. ________	75. ________
16. ________	36. ________	56. ________	76. ________
17. ________	37. ________	57. ________	77. ________
18. ________	38. ________	58. ________	78. ________
19. ________	39. ________	59. ________	79. ________
20. ________	40. ________	60. ________	80. ________

2014 Countdown Round Problems

1. A fair die is rolled. What is the probability that the number 2 is rolled? Express your answer as a common fraction.

2. An equilateral triangle has perimeter 27 units. What is the length of one of its sides?

3. What is the value of $2^7 - 2^6$?

4. For a recipe, Jacob needs 3 gallons of milk or n quarts of milk. What is the value of n?

5. Two dice are rolled. What is the probability that the sum of the numbers rolled is 1? Express your answer in simplest form.

6. Find the mean of the set: {90, 87, 93}.

7. How many quarters are needed to exchange for 425 pennies?

8. If a square has area 49 units^2, what is its perimeter?

9. Bernard sees a movie that runs for 92 minutes. If he starts watching the movie at 1:00 PM, what time will the movie end?

10. What is the sum of all the prime numbers between 1 and 10?

11. What is the Least Common Multiple (LCM) of 6 and 10?

12. Find the mode of the set: {20, 20, 21, 19, 22, 19, 21, 21}.

13. How many yards are equivalent to 417 feet?

14. What is the units digit of the decimal representation of 3^7?

15. What is the Greatest Common Factor (GCF) of 15 and 11?

16. The numerical value of a square's perimeter and area are equal. What is the measurement of its side length?

17. Find the value of $1 - 2 + 3 - 4 + 5 - 6 + 7 - 8 + 9 - 10$.

18. What is the positive difference between 13 yards and 40 feet? Express your answer in inches.

19. April 5, 2014 was on a Saturday. What day of the week was April 25, 2014?

20. If $a + b = 6$ and $a - b = 4$, what is the value of a^2 ?

21. There are 9 total bicycles and tricycles in a parking lot. There are 21 wheels total. How many bicycles are there?

22. How many diagonals does a square have?

23. There are 2 blue marbles and 3 red marbles in a bag. If one marble is drawn at random, what is the probability that it is not blue? Express your answer as a common fraction.

24. How many factors does 45 have?

25. How many ways can the letters in the word "BAT" be arranged?

26. What is the largest number less than 100 that is divisible by 5?

27. Shannon loves the number 5 so much that she finds the product of $5 \times 5 \times 5 \times 5$. What product does she calculate?

28. The probability that it rains on any given day is 10%. What is the probability that it does not rain on Tuesday? Express your answer as a common fraction.

29. Johnny writes down all the digits in the numbers 1 to 15, inclusive. How many digits does Johnny write down in total?

30. How many cents are in 1 quarter, 2 dimes, 3 nickels and 4 pennies?

31. Paul sees a movie that is x minutes long. If the movie starts at 3:25 PM and ends at 5:05 PM, what is the value of x?

32. Julia's car costs $5.6 million while Rebecca's car costs $28.5 million. How much more does Rebecca's car cost than Julia's car in dollars?

33. Find the positive difference between 4102 and 2014.

34. What is the greatest prime factor of 303?

35. What is the perimeter of a rectangle with area 19 if all of the side lengths are positive integers?

36. What is 10 increased by 200% ?

37. Find the range of the set: {3, 1, 4, 0, 9, 10, 5, 7, 8, 2, 6}

38. What is the side length of a cube with volume 125 units^3?

39. How many seconds are in 15 minutes?

40. What is the area of a triangle with sides 3, 4 and 5?

41. What is the measure of one of the angles in an equilateral triangle? Express your answer in degrees.

42. How many days are in 192 hours?

43. Find the circumference of a circle with area 4π. Express your answer in terms of π.

44. What is the remainder when $3 \times 5 \times 7$ is divided by two?

45. What is 20% of 60?

46. Express ⅜ as a decimal.

47. What is the sum of the angles of a triangle? Express your answer in degrees.

48. Solve. $6 \times 5 \times 4 \times 3 \times 2 \times 1$.

49. What is the hypotenuse of a right triangle with legs 5 and 12?

50. How many ounces are in two pounds?

51. A toy car costs 95 cents. What is the most number of toy cars that Todd can buy with his $5.00 allowance?

52. Sally can build eight cars in one hour. How many cars can she build in 30 minutes?

53. Solve. 2 + 4 + 6 + 8 + 10 + 12 + 14 + 16 + 18 + 20.

54. Find the product of all of the digits in the number 125,783,401,839,824.

55. A fair die is rolled. What is the probability that the number rolled is a multiple of two? Express your answer as a common fraction.

56. A bus in Mathville leaves the station every 11 minutes. If a bus just left the station at 10:54 PM, what time will the next bus leave the station?

57. What is the Least Common Multiple (LCM) of 1, 2, and 3?

58. A customer purchases a newspaper for 99 cents with $1.50. What amount of change should the customer receive? Express your answer in cents.

59. How many seconds are in 2 hours?

60. What is the smallest prime factor of 24?

61. Two coins are flipped. What is the probability that zero heads show up? Express your answer as a common fraction.

62. 10 yards is equivalent to x inches. What is the value of x?

63. Find the largest counting number less than $8 \times 6 \times 4 \times 2$.

64. Sarah makes $20 an hour working at a convenience store. If she works 80 hours a month, how much money does she make each month? Express your answer in dollars.

65. How many diagonals does a convex quadrilateral have?

66. Find the units digit of 6^6.

67. Find the value of $20^2 - 14^2$.

68. James took a lunch break from work. He left work at 11:00 AM and returned at 12:37 PM. How long was James' lunch break in minutes?

69. What is the diagonal of a rectangle with side lengths 9 and 12?

70. How many prime numbers are between 30 and 40?

71. How many factors does the number 25 have?

72. A die is rolled. What is the probability that the number rolled is a 5? Express your answer as a common fraction.

73. How many feet are in 5 miles?

74. Express $\frac{3}{5}$ as a percent.

75. What is the value of a if $(a - 2014) = 2014$?

76. Find the surface area of a cube with side length 4.

77. A deluxe toy costs $15.49, while a premium toy costs $29.79. How much more does the premium toy cost than the deluxe toy? Express your answer in dollars.

78. How many pints are in 3 quarts?

79. Find the median of the set: {9, 12, 13, 5, 1}.

80. Solve. $2014 \times 2 - 1007 \times 4$.

2015
Maui Math
Challenge

[This page is intentionally left blank]

MAUI MATH CHALLENGE

2015
■ Individual Round ■
Problems 1-30

Name ____

School Name ______________________________ Grade __________

DO NOT BEGIN UNTIL YOU ARE INSTRUCTED TO DO SO.

1. The Individual Round consists of 20 multiple choice problems and 10 short answer problems.

2. You will have **40 minutes** to complete them.

3. Each Multiple-Choice Problem is worth 1 point and each Short Answer Problem is worth 2 points. There is no partial credit. The maximum score is 40 points.

Multiple Choice	Short Answer	Total Score	Scorer's Initials
$+$	$\times 2$	$=$	
$+$	$\times 2$	$=$	

2015 Individual Round Answer Sheet

1.	(a)	(b)	(c)	(d)	(e)	21. ______
2.	(a)	(b)	(c)	(d)	(e)	
3.	(a)	(b)	(c)	(d)	(e)	22. ______
4.	(a)	(b)	(c)	(d)	(e)	
5.	(a)	(b)	(c)	(d)	(e)	23. ______
6.	(a)	(b)	(c)	(d)	(e)	
7.	(a)	(b)	(c)	(d)	(e)	24. ______
8.	(a)	(b)	(c)	(d)	(e)	
9.	(a)	(b)	(c)	(d)	(e)	25. ______
10.	(a)	(b)	(c)	(d)	(e)	
11.	(a)	(b)	(c)	(d)	(e)	26. ______
12.	(a)	(b)	(c)	(d)	(e)	
13.	(a)	(b)	(c)	(d)	(e)	27. ______
14.	(a)	(b)	(c)	(d)	(e)	
15.	(a)	(b)	(c)	(d)	(e)	28. ______
16.	(a)	(b)	(c)	(d)	(e)	
17.	(a)	(b)	(c)	(d)	(e)	29. ______
18.	(a)	(b)	(c)	(d)	(e)	
19.	(a)	(b)	(c)	(d)	(e)	30. ______
20.	(a)	(b)	(c)	(d)	(e)	

1. Find the 17th positive multiple of 3.

a) 45 b) 48 c) 51 d) 54 e) 57

2. What is the volume of a cube with side length 3 feet?

a) 9 ft^3 b) 54 ft^3 c) 27 ft^3 d) 18 ft^3
e) 81 ft^3

3. Find the next number in the arithmetic sequence: {1, 5, 9, 13, 17, 21, 25, __}

a) 26 b) 27 c) 28 d) 29 e) 30

4. A bag contains 1 red, 1 green and 1 blue marble. Jane randomly selects one marble from the bag. What is the probability that Jane draws a non-red marble?

a) 1 b) ⅔ c) ½ d) ⅓ e) 0

5. If Jack wrote all of the numbers from 1 to 30, how many times would he write the digit 2?

a) 3 b) 10 c) 12 d) 13 e) 14

6. A square has perimeter 28 inches. What is the area of the square?

a) 14 in^2 b) 28 in^2 c) 36 in^2 d) 40 in^2
e) 49 in^2

7. Find the sum of the factors of 14.

a) 14 b) 15 c) 17 d) 24 e) 28

8. Given the set: {1, 11, 111, 1111, 11111}. Find the median.

a) 1 b) 11 c) 61 d) 111 e) 661

9. How many numbers between 1 and 71 are multiples of 5?

a) 12 b) 13 c) 14 d) 15 e) 16

10. How many weeks are in 133 days?

a) 15 b) 15 c) 17 d) 18 e) 19

11. The ratio of boys to girls in a classroom is 1:2. If there are 15 boys in the classroom, how many girls are there?

a) 45 b) 30 c) 20 d) 10 e) 8

12. One newspaper and two magazines cost $11. Two newspapers and one magazine cost $10. How much does one newspaper and one magazine cost?

a) $7 b) $8 c) $9 d) $10 e) $5

13. Find the value of a if $a + b = 16$, $ab = 55$, and $a < b$, where a and b are counting numbers.

a) 5 b) 6 c) 9 d) 10 e) 11

14. What is the area of a rectangle with length 8 meters and diagonal 10 meters?

a) 24 meters^2 b) 48 meters^2 c) 80 meters^2 d) 36 meters^2
e) 96 meters^2

15. Three fair coins are flipped. What is the probability that more heads come up than tails?

a) ⅛ b) ⅜ c) ½ d) ¼ e) ¾

16. Using only pennies, nickels, dimes and quarters, what is the least number of coins necessary to pay a $1.44 toll?

a) 10 b) 11 c) 12 d) 13 e) 14

17. How many minutes are in 3 days?

a) 4320 minutes b) 2160 minutes c) 180 minutes d) 144 minutes
e) 72 minutes

18. A circle has area 144π. What is the positive difference of the numerical value between the area and the circumference of the circle?

a) 12π b) 24π c) 120π d) 132π
e) 144π

19. How many prime numbers are between 23 and 53, inclusive?

a) 5 b) 6 c) 7 d) 8 e) 9

20. Find the number of factors for 48.

a) 8 b) 9 c) 10 d) 11 e) 36

21. Given a set: $\{1, 13, 25, 37, 49, 61, 73, 85\}$. Find the mean.

22. A rectangle has length and width that are whole numbers. If the area of the rectangle is 11 square feet, what is its perimeter? Express your answer in feet.

23. What is the surface area of a rectangular prism with length 6, width 5, and height 4?

24. Patty was giving away a cake with 24 equal pieces. Amanda ate half of the cake. Soon after, Billy ate one third of the remainder of the cake. Finally, Charlie ate the rest of the cake. How many pieces did Charlie eat?

25. Joel drives at a speed of 100 miles per hour. How many miles will he drive in 180 minutes?

26. What is the Least Common Multiple (LCM) of 14 and 35?

27. A cube has volume 27 units^3. What is the surface area of the cube?

28. How many diagonals does a hexagon have?

29. Rocky and Halle are paying their bill at Maui Math Restaurant. They also have to pay a tip that is 15% of the original bill. If the total amount including tip came to $11.50, how much was the tip?

30. What is the 10th smallest prime number?

MAUI MATH CHALLENGE

2015
■ Team Round ■
Problems 1-10

Name ______________________________ Grade ____________

Name ______________________________ Grade ____________

Name ______________________________ Grade ____________

Name ______________________________ Grade ____________

School Name ______________________________

DO NOT BEGIN UNTIL YOU ARE INSTRUCTED TO DO SO.

1. The Team Round consists of 10 short answer problems.

2. You will have **20 minutes** to complete them.

3. Each problem is worth 2 points toward your team score. There is no partial credit. The maximum score on the Team Round is 20 points.

Total Correct	Total Score	Scorer's Initials
$\times 2$	$=$	
$\times 2$	$=$	

2015 Team Round Answer Sheet

1. ______________________________

2. ______________________________

3. ______________________________

4. ______________________________

5. ______________________________

6. ______________________________

7. ______________________________

8. ______________________________

9. ______________________________

10. ______________________________

2015 Team Round Problems

1. If a + b + c = 21, a + b = 14 and a + c = 15, what is the product of b and c?

2. A list of three different positive integers is written in numerical order from least to greatest. The sum of the numbers is 90 and the median is 2. What is the greatest number in the list?

3. A bag has 2 red marbles, 5 blue marbles, 7 green marbles and 11 black marbles. What is the least number of marbles that Troy has to take out to guarantee that he chose at least 1 blue marble?

4. Using only quarters, dimes, and nickels, how many different ways are there to make 40 cents? (Hint: You do not need to use all of the different coins.)

5. A number is randomly chosen from 1 to 100, inclusive. What is the probability that the number is divisible by 2 and 5? Express your answer as a percent.

6. How many ways can the letters in the word MAMMA be arranged?

7. A circle is inscribed in a square with side length 10 units. Find the area of the circle in terms of π.

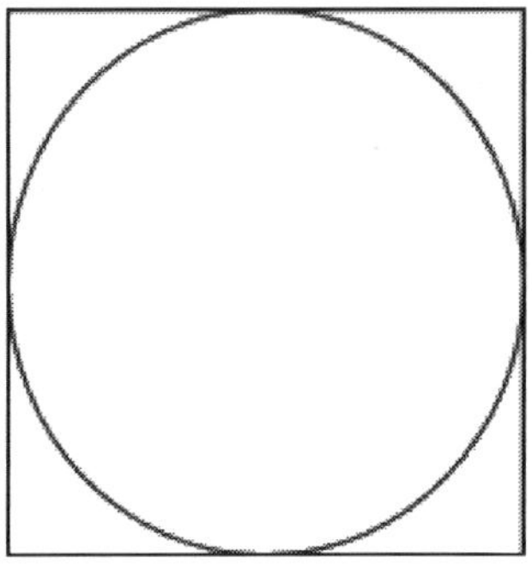

8. The ratio of boys to girls in a classroom is 9 to 1. However, after 20 boys leave the room, the new ratio of boys to girls is 7 to 1. How many total students were there in the beginning?

9. Find the sum of the prime numbers between 20 and 50.

10. Andy, Bill, Carrie, David, Eddy, Freddie, Greg, Hannah and Isabella are at a party. At the beginning of the party, they start to shake hands with each other. If each person shook hands with another person only once, how many handshakes were there in total?

MAUI MATH CHALLENGE

2015
■ Countdown Round ■
Problems 1-80

---

1. The Countdown Round consists of 80 short answer problems.
2. You will have **1 minute** to complete each problem.
3. Each problem is worth 1 point.

2015 Countdown Round Answer Sheet

1. ________	21. ________	41. ________	61. ________
2. ________	22. ________	42. ________	62. ________
3. ________	23. ________	43. ________	63. ________
4. ________	24. ________	44. ________	64. ________
5. ________	25. ________	45. ________	65. ________
6. ________	26. ________	46. ________	66. ________
7. ________	27. ________	47. ________	67. ________
8. ________	28. ________	48. ________	68. ________
9. ________	29. ________	49. ________	69. ________
10. ________	30. ________	50. ________	70. ________
11. ________	31. ________	51. ________	71. ________
12. ________	32. ________	52. ________	72. ________
13. ________	33. ________	53. ________	73. ________
14. ________	34. ________	54. ________	74. ________
15. ________	35. ________	55. ________	75. ________
16. ________	36. ________	56. ________	76. ________
17. ________	37. ________	57. ________	77. ________
18. ________	38. ________	58. ________	78. ________
19. ________	39. ________	59. ________	79. ________
20. ________	40. ________	60. ________	80. ________

2015 Countdown Round Problems

1. How many prime numbers are between 1 and 10?

2. If $a + b = 10$ and a is 7, what is the product of a and b?

3. Find the mode of the set: $\{1, 3, 3, 9, 10\}$.

4. How many pennies are needed to exchange for 20 nickels?

5. Three fair coins are flipped. What is the probability of getting 0 tails? Express your answer as a common fraction.

6. What is the circumference of a circle with area 16π? Express your answer in terms of π.

7. How many integers are between 15 and 87, inclusive?

8. Find the value of $3 + 4 + 5 + 6 + 7 + 8 + 9$.

9. What is 50% of 9134?

10. What is the volume of a cube with side length 8?

11. How many cups are in 1.5 gallons?

12. What is the units digit of 6^4?

13. The number of people in a room was 50. After one hour, 20% of them left the room. What is the remaining number of people in the room?

14. Jack rides his bike at a constant rate of 5 miles per hour. What distance can he travel in one day assuming that he does not stop biking? Express your answer in miles.

15. How many seconds are in 3 hours?

16. Larry bought a toy car at a store for $8.49. What change did he receive if he paid with $10.00? Express your answer in dollars.

17. How many multiples of 10 are between 21 and 133?

18. Find the positive difference between 12345 and 54321.

19. If James makes $10 per hour and Jack makes twice as much as James, how much money does Jack make in 4 hours?

20. Find the mean of the set: $\{1, 2, 3, 4, 5, 6, 7\}$.

21. What is the greatest prime factor of 24?

22. Find the units digit of 5^8.

23. What is the value of $5 \times 4 \times 3 \times 2 \times 1$?

24. Find the sum of the integers from 10 to 15, inclusive.

25. A movie starts at 10:00 PM and ends at 12:31 AM. How long is the movie in minutes?

26. Find the sum of all the factors of 12.

27. What is 10% of 10% of 9800?

28. What is one of the angles in a rectangle? Express your answer in degrees.

29. What is the area of a circle with radius 8? Express your answer in terms of π.

30. Find the value of $1 + 1 - 1 - 1 + 1 + 1 - 1 - 1$.

31. How many diagonals does a rhombus have?

32. How many multiples of 23 are between 48 and 55?

33. Find the value of $5 \times 4 \times 3 \times 2 \times 1 \times 0 + 3 \times 2$.

34. How many dollars are in 5 quarters, 5 nickels and 5 dimes?

35. How many digits does it take to write out all the integers from 1 to 10, inclusive?

36. Find the product: $7 \times 11 \times 13$.

37. If $a \times b = 24$ and $b = 4$, what is the value of $a + b$?

38. What is the hypotenuse of a right triangle with legs 10 and 24?

39. Find the product of all the integers between -15 and 15.

40. If $a = 6$, what is the value of $a^2 - a$?

41. What is the Least Common Multiple (LCM) of 10 and 15?

42. How many yards are in 2 miles?

43. How many ounces are in 3.5 pounds?

44. A regular die is rolled. What is the probability that the number rolled is a multiple of 4? Express your answer as a common fraction.

45. How many ways can the letters in the word "CAR" be arranged?

46. Find the value of $10 - 9 + 8 - 7 + 6 - 5 + 4 - 3 + 2 - 1$.

47. What is the largest prime number less than 100?

48. Express $\frac{2}{5}$ as a percent.

49. What is the value of $2015 + 20 - 15$?

50. Find the positive difference between the smallest prime number and the largest prime number less than 15.

51. What is the area of a square with perimeter 24 feet? Express your answer in square feet.

52. How many minutes are in 7 days?

53. If Jackie buys 3 dolls, each cost 98 cents, with a $5.00 bill, how much should she receive in change? Express your answer in dollars.

54. Find the diagonal of a rectangle with length 8 inches and width 6 inches. Express your answer in inches.

55. What is 20% of 50% of 300?

56. How many integers are between −15 and 24?

57. How many nickels are needed to have the same value as 25 quarters?

58. Find the remainder when 1001 is divided by 13.

59. How many pints are in 5 gallons?

60. Find the greatest multiple of 65 that is less than 200.

61. Find the value of $4 \times 3 \times 2 \times 1 \times 2 \times 3 \times 4$.

62. It takes Julia 45 minutes to drive to work. If work starts at 8:11 AM, what time should she leave her house in order to arrive on time?

63. Two dice are rolled. What is the probability that the sum of the two numbers rolled is 14? Express your answer in simplest form.

64. Find the circumference of a circle with radius 5. Round your answer to the nearest whole number.

65. Find the range of the set: {23, 54, 0, 11, 32, 72, 16, 100, 88, 96}

66. 6 fair coins are flipped. What is the probability that 3 of the coins are heads and 4 of the coins are tails? Express your answer in simplest form.

67. What is the product of the Least Common Multiple (LCM) and the Greatest Common Factor (GCF) of 8 and 12?

68. Find the area of an isosceles right triangle with one leg equal to 12.

69. What is the remainder when $(11 \times 9 \times 7 \times 5 \times 3)$ is divided by 2?

70. Find the value of $5! + 4! + 3! + 2! + 1!$

71. How many ways can the letters in the word "DARE" be arranged?

72. What is the product of the first 10 whole numbers?

73. Find the value of 11 + 13 + 15 + 17 + 19.

74. What is the sum of all the angles in a pentagon? Express your answer in degrees.

75. How many miles are in 36,960 feet?

76. Talia watches five 90-minute movies in rapid succession. If she starts at 12:00 PM, what time will she finish?

77. What is the perimeter of a square with area 400?

78. What is the remainder when 20^{15} is divided by 100?

79. Find the mode of the set: {11, 11, 12, 17, 13, 11, 12, 19, 19, 17, 12, 11}.

80. Find the number of integers from −2015 to 2015, inclusive.

2016
Maui Math Challenge

[This page is intentionally left blank]

MAUI MATH CHALLENGE

2016

■ Individual Round ■

Problems 1-30

Name ____

School Name ______________________________ Grade _________

DO NOT BEGIN UNTIL YOU ARE INSTRUCTED TO DO SO.

1. The Individual Round consists of 20 multiple choice problems and 10 short answer problems.

2. You will have **40 minutes** to complete them.

3. Each Multiple-Choice Problem is worth 1 point and each Short Answer Problem is worth 2 points. There is no partial credit. The maximum score is 40 points.

Multiple Choice	Short Answer	Total Score	Scorer's Initials
+	×2	=	
+	×2	=	

2016 Individual Round Answer Sheet

1.	(a)	(b)	(c)	(d)	(e)	21. ____________
2.	(a)	(b)	(c)	(d)	(e)	
3.	(a)	(b)	(c)	(d)	(e)	22. ____________
4.	(a)	(b)	(c)	(d)	(e)	
5.	(a)	(b)	(c)	(d)	(e)	23. ____________
6.	(a)	(b)	(c)	(d)	(e)	
7.	(a)	(b)	(c)	(d)	(e)	24. ____________
8.	(a)	(b)	(c)	(d)	(e)	
9.	(a)	(b)	(c)	(d)	(e)	25. ____________
10.	(a)	(b)	(c)	(d)	(e)	
11.	(a)	(b)	(c)	(d)	(e)	26. ____________
12.	(a)	(b)	(c)	(d)	(e)	
13.	(a)	(b)	(c)	(d)	(e)	27. ____________
14.	(a)	(b)	(c)	(d)	(e)	
15.	(a)	(b)	(c)	(d)	(e)	28. ____________
16.	(a)	(b)	(c)	(d)	(e)	
17.	(a)	(b)	(c)	(d)	(e)	29. ____________
18.	(a)	(b)	(c)	(d)	(e)	
19.	(a)	(b)	(c)	(d)	(e)	30. ____________
20.	(a)	(b)	(c)	(d)	(e)	

2016 Individual Round Problems

1. John's favorite integer is between 10 and 20, inclusive, and contains the digit 7. What is John's favorite integer?

a) 7 b) 14 c) 17 d) 21 e) 77

2. Find the product of the range and the mode of the set:
$\{0, 100, 23, 2, 64, 23, 44, 65, 78, 12, 12, 2, 2\}$.

a) 100 b) 102 c) 200 d) 1176
e) 2254

3. How many cups are in 6 gallons of water?

a) 24 cups b) 36 cups c) 48 cups d) 96 cups
e) 192 cups

4. The area of a square is equivalent to the area of a rectangle. If the side lengths of the rectangle are 3 units and 12 units, what is the perimeter of the square?

a) 36 units b) 24 units c) 20 units d) 15 units
e) 6 units

5. What is the value of $1 \times 2 + 3 \times 4 + 5 \times 6 + 7 \times 8 + 9 \times 10$?

a) 55 b) 180 c) 190 d) 362880
e) 3628800

6. How many positive factors does 36 have?

a) 6 factors   b) 7 factors   c) 8 factors   d) 9 factors
e) 10 factors

7. A shipping service company packs pots in boxes, then packs boxes in containers. 6 pots fit in 1 box and 8 boxes fit in 1 container. If there are 720 pots to be shipped, what is the least number of containers that are needed?

a) 5 containers   b) 8 containers   c) 10 containers   d) 15 containers
e) 20 containers

8. What is the product of the smallest prime number and the smallest composite number?

a) 2   b) 3   c) 6   d) 8   e) 12

9. Dominic loves to collect pennies, nickels and dimes. If he has 3 of each of these coins, how much money does he have?

a) 16 cents   b) 18 cents   c) 32 cents   d) 33 cents
e) 48 cents

10. Kyle leaves home for work at 7:30 AM and returns home at 6:30 PM. If it takes him 35 minutes to drive from home to work for each trip, how long does he stay at work? (Assuming he does not leave his workplace until he goes home)

a) 11 hrs   b) 10 hrs and 25 min   c) 10 hrs and 10 min   d) 9 hrs and 50 min
e) 9 hrs

11. If a woodchuck could chuck 24 lbs of wood in 6 days, how many days would it take to chuck 52 lbs of wood? (Assume it chucks wood at a constant rate.)

a) 12 days b) 13 days c) 14 days d) 208 days
e) 209 days

12. A rectangle and a square have the same perimeter. If the side lengths of the rectangle are 5 units and 19 units, what is the area of the square?

a) 36 units^2 b) 48 units^2 c) 95 units^2 d) 144 units^2
e) 256 units^2

13. What is the Greatest Common Factor (GCF) of 720 and 168?

a) 2 b) 4 c) 12 d) 24 e) 36

14. Dane is driving at a speed of 55 miles per hour. Assuming he drives at a constant rate, how long does it take him to travel 1,210 miles?

a) 22 hours b) 21 hours c) 19 hours d) 600 minutes
e) 1200 minutes

15. A circle has circumference 12π. If Jane cuts the circle in half, what is the area of the half-circle?

a) 6π b) 12π c) 18π d) 24π e) 36π

16. The ratio of boys to girls in a classroom is 3:5. If there are 24 total students in the class, how many of the students are boys?

a) 3 boys b) 6 boys c) 9 boys d) 15 boys
e) 18 boys

17. How many diagonals does a regular octagon have?

a) 19 b) 20 c) 21 d) 22 e) 28

18. Joe took a number and multiplied it by 2 when he meant to divide it by 2. Then he took that result and added 2 when he meant to subtract 2, ending up with 34. If Joe did what he meant to do, what number would he have ended up with?

a) 32 b) 16 c) 10 d) 8 e) 6

19. Josh flips a coin and rolls a 6-sided die. What is the probability that he flips a heads and rolls a 5?

a) $\frac{1}{24}$ b) $\frac{1}{12}$ c) ⅛ d) ⅙ e) $\frac{5}{24}$

20. One of three sisters, Anna, Bella, and Celina, ate the last cookie from the cookie jar. When asked by their mother, each responded differently:
Anna: I didn't eat it!
Bella: Celina ate it!
Celina: Anna ate it!
If Anna is the only one telling the truth, who ate the last cookie from the cookie jar?

a) Anna b) Bella c) Celina d) Their mother
e) Not Enough Information

21. Find the largest three digit number with distinct digits.

22. What is the positive difference between the numerical value of the volume and the numerical value of the surface area of a cube with side length 10?

23. The ratio of dogs to cats at an animal shelter was 6:4. After 10 dogs were adopted, the ratio of dogs to cats became 1:1. How many cats are there at the animal shelter?

24. Let N be 1,234,567,890,987,654,321. Now, let the product of the digits in N be P and the sum of the digits in N be S. What is the product of P and S?

25. What is the volume of a cube with surface area 24 units^2?

26. Let $a + b + c = 13$ and $abc = 11$. If a, b, and c are positive whole numbers, find the value of $ab + bc + ac$.

27. There are 20 people in a room. If each person shakes hands with everyone else exactly once, how many handshakes take place? (Assume a person does not shake hands with him/herself.)

28. Calculate the sum: $1 + 2 + 4 + 8 + 16 + 32 + 64 + 128 + 256 + 512$

29. Find the area of the following pinwheel which is comprised of a square with side length 12 units and 4 isosceles right triangles.

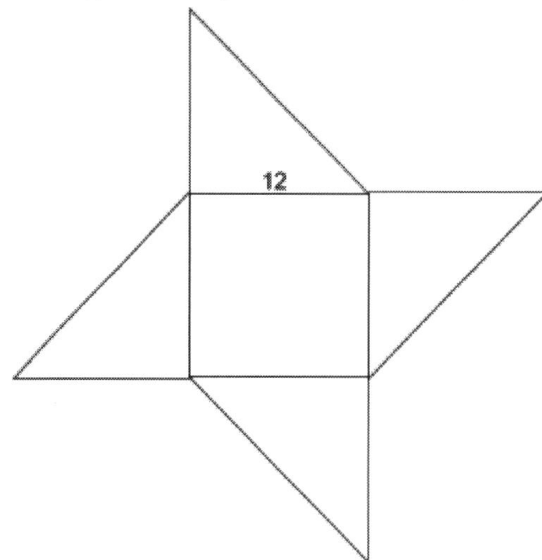

30. Jackie's dad's age is three times Jackie's age. 8 years ago, her dad's age was eleven times Jackie's age then. What is the sum of Jackie's current age and her dad's current age? Express your answer in years.

MAUI MATH CHALLENGE

2016
■ Team Round ■
Problems 1-10

Name ______________________________ Grade ____________

Name ______________________________ Grade ____________

Name ______________________________ Grade ____________

Name ______________________________ Grade ____________

School Name ____

DO NOT BEGIN UNTIL YOU ARE INSTRUCTED TO DO SO.

1. The Team Round consists of 10 short answer problems.
2. You will have **20 minutes** to complete them.
3. Each problem is worth 2 points toward your team score. There is no partial credit. The maximum score on the Team Round is 20 points.

Total Correct	Total Score	Scorer's Initials
$\times 2$	$=$	
$\times 2$	$=$	

2016 Team Round Answer Sheet

1. ______________________

2. ______________________

3. ______________________

4. ______________________

5. ______________________

6. ______________________

7. ______________________

8. ______________________

9. ______________________

10. ______________________

2016 Team Round Problems

1. Given a set: {1003, 560, 77, 880, 990099, 2801, 302, 0, 80, 1120}. If one of the numbers is chosen at random, what is the probability that the product of all of its digits is positive? Express your answer as a common fraction.

2. A regular triangle, a square, a regular pentagon and a regular hexagon all have the same perimeter. If the side length of the pentagon is 24 units, what is the sum of the side length of the triangle, the side length of the square, the side length of the pentagon and the side length of the hexagon?

3. Find the sum of all the two-digit numbers that are equivalent to the original number when written in reverse order. (For example, 121 is written the same forwards and backwards.)

4. What is the remainder when 5^{10} is divided by 100?

5. A sequence of numbers follow a rule that, starting with the third term, the next term is the sum of the previous two terms. The sequence is 6, a, b, 12, c, d. If a, b, c, and d are counting numbers, what is the sum of a, b, c, and d?

6. Johnny owns a store that sells tricycles and bicycles for all ages. In his inventory right now, he has 12 more tricycles than bicycles and there are 161 wheels in total. How many bicycles does Johnny have in his inventory? (Assume tricycles have 3 wheels and bicycles have 2 wheels)

7. Two cars, 225 miles apart, are driving towards each other. If the first car is going twice as fast as the second car, how many miles will the first car travel when they meet? (Assume the cars are moving at a constant rate.)

8. In the figure below, a rectangle holds 12 congruent circles that are tangent to the rectangle's sides and to each other. If the perimeter of the rectangle is 84, what is the total area of the 12 circles? Express your answer in terms of π.

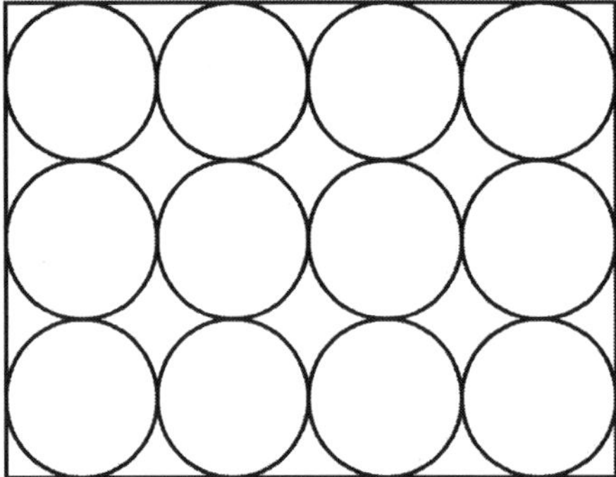

9. A list of 5 positive integers is written in numerical order from the least to the greatest. If the median, mode, and mean is 4, what is the greatest possible value of the largest number in the list?

10. On the top row of a bookshelf, there are 2 math books (Algebra and Geometry) and 3 science books (Biology, Chemistry and Physics). Blake wants to arrange the books in a way where the 2 math books are grouped together and the 3 science books are grouped together. Given that the books can be arranged in their group, how many ways can Blake arrange these 5 books from left to right? (An example is Biology-Chemistry-Physics-Geometry-Algebra.)

MAUI MATH CHALLENGE

2016

■ Countdown Round ■

Problems 1-110

---

1. The Countdown Round consists of 110 short answer problems.

2. You will have **1 minute** to complete each problem.

3. Each problem is worth 1 point.

2016 Countdown Round Answer Sheet

1. ____________
2. ____________
3. ____________
4. ____________
5. ____________
6. ____________
7. ____________
8. ____________
9. ____________
10. ____________
11. ____________
12. ____________
13. ____________
14. ____________
15. ____________
16. ____________
17. ____________
18. ____________
19. ____________
20. ____________
21. ____________
22. ____________
23. ____________
24. ____________
25. ____________
26. ____________
27. ____________
28. ____________
29. ____________
30. ____________
31. ____________
32. ____________
33. ____________
34. ____________
35. ____________
36. ____________
37. ____________
38. ____________
39. ____________
40. ____________
41. ____________
42. ____________
43. ____________
44. ____________
45. ____________
46. ____________
47. ____________
48. ____________
49. ____________
50. ____________
51. ____________
52. ____________
53. ____________
54. ____________
55. ____________
56. ____________
57. ____________
58. ____________
59. ____________
60. ____________

2016 Countdown Round Answer Sheet

61. ____________
62. ____________
63. ____________
64. ____________
65. ____________
66. ____________
67. ____________
68. ____________
69. ____________
70. ____________
71. ____________
72. ____________
73. ____________
74. ____________
75. ____________
76. ____________
77. ____________
78. ____________
79. ____________
80. ____________
81. ____________
82. ____________
83. ____________
84. ____________
85. ____________
86. ____________
87. ____________
88. ____________
89. ____________
90. ____________
91. ____________
92. ____________
93. ____________
94. ____________
95. ____________
96. ____________
97. ____________
98. ____________
99. ____________
100. ____________
101. ____________
102. ____________
103. ____________
104. ____________
105. ____________
106. ____________
107. ____________
108. ____________
109. ____________
110. ____________

2016 Countdown Round Problems

1. How many integers are between 2 and 12, inclusive?

2. What is the product of the number of sides in a rectangle and the number of sides in a hexagon?

3. Find the difference between the two sums, (3 + 5 + 7 + 9 + 11) and (11 + 9 + 7 + 5 + 3).

4. How many pints are in 13 gallons?

5. If A = 199, B = 299 and C = 399, what is the value of C + A + B?

6. A die is rolled. What is the probability that it does not roll a composite or a prime number? Express your answer as a common fraction.

7. Which of these numbers is a prime number? {112, 137, 184, 156}.

8. What is the area of a circle with circumference 22π? Express your answer in terms of π.

9. Eric has 7 pennies, 4 nickels, 3 dimes and 3 quarters. How many cents does he have?

10. Miranda is driving her new car at a constant speed of 31 miles per hour. How long does it take for her to travel 403 miles? Express your answer in hours.

11. How many yards are in 612 inches?

12. Find the value of 25 + 45 + 65 + 85 + 105.

13. How much greater is the perimeter of a square with side length 37 units than the perimeter of a square with side length 22 units?

14. Find the remainder when 123,456,789 is divided by 5.

15. Pablo has 34 chickens and 27 horses on his farm. Assuming chickens have 2 legs and horses have 4 legs, how many animal legs are there in total on the farm?

16. John's birthday is on February 2, which is a Tuesday in 2016. If his younger sister's birthday is on February 29, what day of the week is her birthday in 2016?

17. If three coins are flipped, what is the probability that the outcome is 0 heads?

18. If $a + b = 30$ and $a = 11$, what is the value of $a \times b$?

19. How many positive factors does 30 have?

20. Gabi has $100 to spend at the mall. She first bought a doll for $15 then she saw a movie with popcorn for $27 and lastly she bought a dress for $46. How much money does she have left? Express your answer in dollars.

21. Find the positive difference between the area of a circle with radius 13 and the area of a circle with radius 8. Express your answer in terms of π.

22. A theater is screening a documentary that lasts 87 minutes. If the documentary starts at 6:49 PM, what time will it finish?

23. Find the mean of the set: {3, 11, 19, 27, 35}.

24. David is typing at a speed of 74 words per minute. How long does it take him to type 111 words? Express your answer in seconds.

25. How many positive factors does the number 91 have?

26. Find the value of $5^2 - 4^2 + 3^2 - 2^2 + 1^2$.

27. Kaitlyn needs to buy 12 gallons of water at the store. How many cups of water does she need?

28. What is the Greatest Common Factor (GCF) of 64 and 48?

29. How many minutes are there in five days?

30. Find the remainder when 174,357,818 is divided by 9.

31. How many diagonals does a rectangle have?

32. Lauren is adding all the integers from -13 to 14, inclusive. What is her sum?

33. What is 5% of 20% of 500?

34. What is the largest four-digit number divisible by 5?

35. How many ways can the letters in the name "ALANA" be arranged?

36. Kady has 234 quarters. If a gumball costs $1, how many gumballs can she buy?

37. How many seconds are in 8 hours?

38. If the area of a square is 225 units^2, what is its perimeter?

39. What is the sum of all the multiples of 7 between 1 and 40?

40. A bag contains 5 red marbles, 7 blue marbles and 9 green marbles. If one marble is chosen at random out of the bag, what is the probability that it is a blue marble? Express your answer as a common fraction.

41. There are 17 cars and 12 motorcycles in a parking lot. How many wheels are there in total?

42. McKenna drove for 15 hours and traveled 825 miles. On average, how fast was she driving in miles per hour?

43. A high school's student council has 4 members: Aisha, Braden, Chiemi and Darla. For a fundraiser, they need 2 members of student council to run it. How many different pairs of members can be selected to run the fundraiser?

44. Find the positive difference between the total number of sides of 33 pentagons and the total number of sides of 34 octagons.

45. How many integers between 1 and 50, inclusive, contain the digit 1?

46. Poe is flying in his cruiser at a speed of 1370 miles per hour. How many hours will it take him to travel 9590 miles?

47. How many positive factors does the number 101 have?

48. A basketball player's height is 7 feet and 6 inches. What is his height in inches?

49. Find the median of this set: {–113, –99, 16, 68, 1234, 2016}.

50. Melanie took a break from work for 113 minutes. If she returned to work at 4:34 PM, what time did her break start?

51. Steven is thinking of a whole number greater than 30 but less than 40 that is divisible by 5. What number is Steven thinking of?

52. Find the positive difference between the numerical value of the circumference of a circle and the numerical value of the area of a circle with radius 14. Express your answer in terms of π.

53. Carver is going to the movies and brings $18.79. If the movie ticket costs $10.48 and popcorn costs $5.93, how much money will he have left after paying for these two items? Express your answer in dollars.

54. Gauss loves to add numbers. If he adds all the integers from 1 to 24, inclusive, what sum will he have?

55. Josh has a sock drawer that has 4 black socks, 3 yellow socks and 5 purple socks. What is the least number of socks that he must take out to guarantee that he has drawn at least 1 black sock?

56. How many positive factors does the number 28 have?

57. If a die is rolled, what is the probability that a multiple of 7 is rolled? Express your answer in simplest form.

58. How many teaspoons are in 10 tablespoons?

59. Find the area of a square with perimeter 64 units.

60. Find the units digit of 2016^2.

61. What is the largest multiple of 11 less than 100?

62. Two dice are rolled. What is the probability that the sum of the numbers rolled is 2? Express your answer as a common fraction.

63. Bethany adds up the first 20 odd natural numbers while Clark adds up the first 20 even natural numbers. What is the positive difference between these two sums?

64. A square's side length and a circle's radius are equivalent in length. If the circumference of the circle is 10π, what is the square's area?

65. Find the value of $(2^0 + 16) \times (20^1 - 6)$.

66. How many multiples of 100 are between 1 and 2016?

67. How many diagonals does a regular nonagon have?

68. How many whole numbers are between -879 and 100?

69. Lucy is trying to figure out Susan's favorite number. If her favorite number is a three-digit positive integer that is divisible by 7 and 100, what is her favorite number?

70. What is the measure of one of the angles in a regular hexagon? Express your answer in degrees.

71. Find the value of $1^2 + 2^2 + 3^2 + 4^2 + 5^2$.

72. Albert can clap 7 times per second. If he continues at this rate, how many times can he clap in an hour?

73. What is the Least Common Multiple (LCM) of 2 and 1001?

74. How many quarters are needed to exchange for $15.75?

75. A decagon and a heptagon have the same perimeter. If the side length of the decagon is 21 units, what is the side length of the heptagon?

76. Find the sum of the factors of 16.

77. A car weighs exactly 2.125 tons. How many pounds does this car weigh?

78. Zofia runs at a constant rate of 5 feet per second. How many minutes will it take her to travel 400 yards?

79. A bag contains 2 blue marbles and 3 red marbles. What is the least number of marbles that must be pulled out of the bag to guarantee that at least 1 blue marble is drawn?

80. Find the value of $20^2 + 20 - 16^2 - 16$.

81. Joshua, Jessica, Jonathan and Jedidiah are sitting in a car. If there are four different seats including the driver's seat and Jonathan must drive the car, how many ways can the four people sit in the car?

82. How many whole numbers are between 11 and 1000, inclusive?

83. Find the side length of a regular nonagon with perimeter 144.

84. Find the median of the set: {1, 33, 999, 12, 16, 0}.

85. Jacob read a 300-page book over the period of three days. If he read 87 pages on the first day and 123 pages on the second day, how many pages did he read on the third day?

86. What is the remainder when 375,123,898 is divided by 8?

87. Robert wants to buy a pack of gum from a vending machine which costs $1.85. What is the number of nickels that Robert needs to pay for the gum?

88. Find the value of $(1^1 + 2^2 + 3^3 + 4^4)$.

89. A square's side length is equivalent to the diameter of a circle. If the square's perimeter is 400, what is the area of the circle? Express your answer in terms of π.

90. What is the largest prime factor of 108?

91. Consider all the arrangements of the letters in the word "ACE." What fraction of the arrangements start with the letter "A"? Express your answer as a common fraction.

92. Chris can type 1200 words in 15 minutes. On average, how many words can he type per minute?

93. Jake has 20 toy cars. If Julia has three times as many toy cars as Jake and Jack has five times as many toy cars as Julia, how many toy cars does Jack have?

94. Find the other leg of a right triangle with area 100 and leg 8.

95. In a barn, the ratio of apples to pears is 1 to 2. If there are 68 apples in the barn, how many pears are there?

96. Lucy has 12 different colored dresses and 11 different colored bows. How many different dress-bow combinations can Lucy make if she does not want to wear her red bow?

97. What is the product of all the whole numbers between 0 and 6, inclusive?

98. Joseph flies an airplane at a constant speed of 600 miles per hour. How many minutes will it take him to travel 450 miles?

99. If an integer is chosen at random between 1 and 10, inclusive, what is the probability that the integer contains the digit 1? Express your answer as a percent.

100. Find the value of $(101^2 - 99^2)$.

101. Carlos has a one square foot wooden board. If a toy car takes up two square inches of space, what is the least number of toy cars that can fill the wooden board's area?

102. Find the numerical sum of the circumference and the area of a circle with diameter 40. Express your answer in terms of π.

103. Jordan wants to find the cube of a number with her calculator. Instead of finding the cube, she accidentally calculated the perfect square of the positive number and got 16. What number should she have gotten if she found the cube of the number?

104. How many multiples of 7 are between 70 and 700, inclusive?

105. Adam bought three gallons of milk. If he drinks three pints of milk a day, how many days will the three gallons of milk last?

106. What is the positive difference between the sum of all the even numbers and the sum of all the odd numbers between 1 and 20, inclusive?

107. A square and a regular pentagon has the same perimeter. If the side length of the pentagon is 4, what is the area of the square?

108. A farm has 24 chickens and 15 cows. If chickens have two legs and cows have four legs, how many animal legs are there in total?

109. Gerald rolls two fair dice. What is the probability that the sum of the two numbers rolled is 11 or 12? Express your answer as a common fraction.

110. James was born on July 1 which was a Saturday. His best friend, Justin was born on July 31 the same year as James. What day of the week was Justin's birth date?

2013 Answer Keys and Solutions

2013 Individual Round Solutions

Answer Key

1. a) 24
2. d) 8
3. e) 1
4. d) 27
5. d) 154
6. a) ½
7. c) 27
8. c) 78,512
9. e) 66
10. d) 16 minutes
11. c) 84 boys
12. b) 5 diagonals
13. a) $25 per hour
14. e) ⅞
15. b) 9 hours
16. d) 20 coins
17. b) 240
18. b) 60
19. b) ⅜
20. b) 455
21. 8 (pencils)
22. 120
23. 21 (outfits)
24. 0
25. 44 (buses)
26. 15 (numbers)
27. 4
28. 12 (ways)
29. 45 (handshakes)
30. 5

1. What is the perimeter of a square with side length 6?

Answer: a) 24

Solution: The formula for calculating the perimeter of a square is P = 4s.
s = 6
P = 4s = 4 × 6 = 24

2. Find the perfect cube of the first prime number.

Answer: d) 8

Solution: The first prime number is 2 because it is the smallest positive integer with two factors: 1 and 2. The perfect cube would be $2^3 = 2 \times 2 \times 2 = 8$.

3. What is the positive difference between 2^3 and 3^2?

Answer: e) 1

Solution: $2^3 = 2 \times 2 \times 2 = 8$ and $3^2 = 3 \times 3 = 9$. The positive difference is 9 – 8 = 1.

4. Find the mean of the set: {22, 22, 28, 29, 34}.

Answer: d) 27

Solution: There are 5 numbers.
The mean is calculated by dividing the sum of the 5 numbers by 5.

$$\text{Mean} = \frac{22 + 22 + 28 + 29 + 34}{5} = \frac{135}{5} = 27$$

5. Using the approximation $\pi = \frac{22}{7}$, what is the area of a circle with radius 7?

Answer: d) 154

Solution: The formula for the area of a circle is $A = \pi r^2$ where r is the radius of the circle.
$r = 7$.

$$\text{Area} = \pi r^2 = \frac{22}{7} * 7^2 = 22 * 7 = 154$$

6. A regular die is rolled. What is the probability that a prime number is rolled?

Answer: a) ½

Solution: A regular die has sides: 1, 2, 3, 4, 5, 6. Out of these six numbers, only three numbers (2, 3, and 5) are prime numbers because each of these numbers has two factors: one and itself.

Thus, there are 3 favorable outcomes and 6 possible outcomes: $P = \frac{3}{6} = \frac{1}{2}$

7. How many quarters are needed to pay for a $6.75 toll?

Answer: c) 27

Solution: A quarter has a value of 25 cents. $6.75 = 675 cents.
675 cents ÷ 25 cents = 27 quarters

8. Which of the following numbers is divisible by 8?

Answer: c) 78,512

Solution: The divisibility rule for 8 is if the last three digits of the number is divisible by 8, the number is divisible by 8.
412, 612, 388, and 122 are not divisible by 8, whereas 512 is divisible by 8. Thus, 78,512 is divisible by 8.

9. Find the sum: 1 + 2 + 3 + … + 9 + 10 + 11.

Answer: e) 66

Solution: The formula for finding the sum of an arithmetic sequence is: $S_n = \frac{n(a_1 + a_n)}{2}$
(See Formula and Tips)
$n = 11, a_1 = 1, a_{11} = 11$

$S_n = \frac{(1+11)*11}{2} = \frac{12*11}{2} = 6*11 = 66$

10. Joe has 35 minutes for his lunch break. He left his office from 10:17 AM to 11:08 AM. How many minutes did he go over his lunch break?

Answer: d) 16 minutes

Solution: There are 51 minutes from 10:17 AM to 11:08 AM.
51 minutes – 35 minutes = 16 minutes.

11. The ratio of boys to girls in a classroom is 7:5. If there are 60 girls in the classroom, how many boys are there?

Answer: c) 84 boys

Solution: There are 7 parts that make up the boys and 5 parts that make up the girls.
60 ÷ 5 = 12. 12 × 7 = 84.

12. How many diagonals does a regular pentagon have?

Answer: b) 5 diagonals

Solution: The formula for the number of diagonals in a convex n-gon is: $d = \frac{n(n-3)}{2}$
(See Formula and Tips)

Thus, $d = \frac{5(5-3)}{2} = 5$

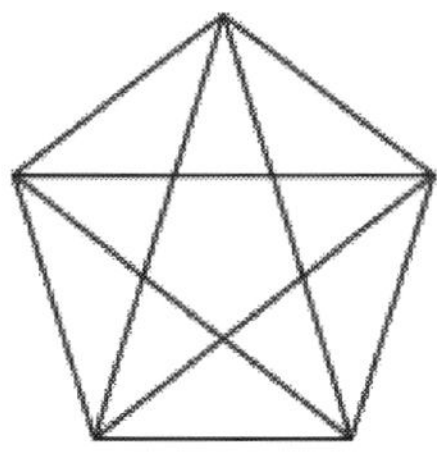

13. The President of a given planet gets paid $400,000 a year. If this planet has 20 months per year, 25 days per month and 32 hours per day, how much does the President get paid per hour?

Answer: a) $25 per hour

Solution:

$$\frac{400000 \text{ dollars}}{1 \text{ year}} * \frac{1 \text{ year}}{20 \text{ months}} * \frac{1 \text{ month}}{25 \text{ days}} * \frac{1 \text{ day}}{32 \text{ hours}} = \frac{400000}{20 * 25 * 32} = \frac{2^7 * 5^7}{2^7 * 5^5} =$$
$$5^2 = 25 \text{ dollars per hour}$$

14. The probability of having a boy or a girl is equally likely. A family has three children. What is the probability that the family has at least one boy?

Answer: e) ⅞

Solution: To find the probability that the family has at least one boy, find the probability that the family has no boys and subtract that probability from 1 (Complementary Counting).

$$\text{P(0 Boys)} = \text{P(Girl)} * \text{P(Girl)} * \text{P(Girl)} = \frac{1}{2} * \frac{1}{2} * \frac{1}{2} = \frac{1}{8}$$
$$\text{P(At Least One Boy)} = 1 - \text{P(0 Boys)} = 1 - \frac{1}{8} = \frac{7}{8}$$

15. Laura and Lucy live 540 miles apart. They both leave their house at the same time and drive toward each other. If Laura is driving at a rate of 45 miles per hour and Lucy is driving at a rate of 15 miles per hour, how many hours will it take for them to meet?

Answer: b) 9 hours

Solution: If they are driving toward each other, the total speed that they are traveling is (15 + 45) = 60 miles per hour.

Using the formula: Distance = Rate × Time
Time = Distance ÷ Rate = (540 miles) ÷ (60 miles per hour) = 9 hours

16. Using only pennies and quarters, what is the least number of coins necessary to make 68 cents?

Answer: d) 20 coins

Solution: To minimize the number of coins, the coins with the largest value should be maximized. Since the penny is worth 1 cent and the quarter is worth 25 cents, the quarter should be maximized.

68 ÷ 25 = 2 R 18. Thus, 2 quarters are used and 18 pennies are used to make up the remaining 18 cents. Hence, 2 + 18 = 20 coins.

17. What is the value of $32^2 - 28^2$?

Answer: b) 240

Solution: Using the formula for the difference of two squares $a^2 - b^2 = (a - b)(a + b)$,
$32^2 - 28^2 = (32 - 28)(32 + 28) = 4 \times 60 = 240$

18. A right triangle has leg 8 and hypotenuse 17. What is the area of the triangle?

Answer: b) 60

Solution: The pythagorean triple that matches this right triangle is 8-15-17 (See Formula and Tips). The other leg can also be found by using the pythagorean theorem and the difference of two squares:

$$\sqrt{17^2 - 8^2} = \sqrt{(17-8)(17+8)} = \sqrt{9 * 25} = 3 * 5 = 15$$

$l_1 = 8$ and $l_2 = 15$:

Thus, the area of the right triangle is $\dfrac{l_1 l_2}{2} = \dfrac{8 * 15}{2} = 60$

19. A coin is flipped three times. What is the probability that exactly two of the coins are heads?

Answer: b) ⅜

Solution: There are 2^3 or 8 total possibilities for 3 coin flips.
HHT, HTH, and THH are the only 3 possibilities that has exactly 2 heads.
Thus, the probability is ⅜.

20. What is the Least Common Multiple (LCM) of 65 and 91?

Answer: b) 455

Solution: Find the prime factorization for each number: $65 = 5^1 13^1$ and $91 = 7^1 13^1$.
Both numbers share the factor 13.

The Least Common Multiple will contain one factor of 5, one factor of 7 and one factor of 13 so that it will be divisible by both numbers: $5^1 7^1 13^1 = 455$.

21. John has three more than half of the number of pencils that James has. If James has 10 pencils, how many pencils does John have?

Answer: 8 (pencils)

Solution: $\text{John} = 3 + \frac{\text{James}}{2}$

Knowing that James' pencils = 10, the number of pencils that John has can be found:

$$\text{John} = 3 + \frac{10}{2} = 3 + 5 = 8$$

22. What is the product of the first 5 natural numbers?

Answer: 120

Solution: The first 5 natural numbers are 1, 2, 3, 4 and 5.
Thus, the product is 5 × 4 × 3 × 2 × 1 = 5! = 120

23. Anna's outfit consists of one shirt and one pair of pants. If she has 3 different pairs of pants and 7 different shirts, how many combinations of outfits does she have?

Answer: 21 (outfits)

Solution: Using the Fundamental Theorem of Counting, the number of combinations of outfits is: Pants × Shirts = 3 × 7 = 21.

24. If the first term of an arithmetic sequence is 0, the second term is 5 and the third term is 10, what is the product of the first term and the 2013th term?

Answer: 0

Solution: The first term is 0. Since the product of 0 and any number is 0, the product of the first term and the 2013th term is 0.

25. Each school bus has 23 passenger seats. If 1,000 students go on a field trip and ride the school bus, what is the least number of buses needed?

Answer: 44 (buses)

Solution: 1000 ÷ 23 = 43 R 11
Thus, 44 buses are needed.

26. How many prime numbers are between 1 and 50?

Answer: 15 (numbers)

Solution: The prime numbers between 1 and 50 are:
2, 3, 5, 7, 11, 13, 17, 19, 23, 29, 31, 37, 41, 43, 47

This totals to 15 numbers.

27. Let $n! = n \times (n-1) \times (n-2) \times \ldots \times 3 \times 2 \times 1$. For example, $4! = 4 \times 3 \times 2 \times 1 = 24$. What is the tens digit of $7!$?

Answer: 4

Solution: $7! = 7 \times 6 \times 5 \times 4 \times 3 \times 2 \times 1 = 5040$
The tens digit is 4.

28. How many ways can the letters in the word "BOOT" be arranged?

Answer: 12 (ways)

Solution: There are four letters which can be arranged 4! ways. However, the two O's can be arranged 2! ways. Thus, the answer is $\frac{4!}{2!} = \frac{24}{2} = 12$.

29. There are ten people in a room. If each person shakes hands with everyone else once, how many handshakes are there? (Note: A person does not shake hands with him/herself)

Answer: 45 (handshakes)

Solution: Each person shakes hand with 9 other people ($10 - 1 = 9$).
Thus, $10 \times 9 = 90$. However, this number must be divided by two because the first person shaking hands with the second person is the same as the second person shaking hands with the first person.

Therefore, $90 \div 2 = 45$ handshakes

30. The set {a, b, c} is given such that: $a < b < c$. The sums $(a + b)$, $(a + c)$, and $(b + c)$ are equal to 10, 11, and 15, respectively. What is the value of $(c - a)$?

Answer: 5

Solution: Subtracting $(a + b = 10)$ from $(b + c = 15)$ will yield the answer:

$$\begin{array}{l} \quad b + c = 15 \\ \underline{-\ (a + b = 10)} \\ \quad c - a = 5 \end{array}$$

(The values of a, b, and c are 3, 7, and 8, respectively)

2013 Team Round Solutions

Answer Key

1. 1
2. $\frac{5}{17}$
3. 88
4. 41
5. 7 (minutes)
6. 72 (sandwiches)
7. 61
8. 55
9. 50
10. 30 (ways)

1. What is the remainder when 5^6 is divided by 2?

Answer: 1

Solution: $5^6 = 5 \times 5 \times 5 \times 5 \times 5 \times 5$ which is the product of 6 odd numbers. The product of an odd number and another odd number is an odd number (See Formula and Tips). Thus, the product of 6 odd numbers will be an odd number.

An odd number will always have a remainder of 1 when divided by 2.

2. All of the 17 letters in MAUI MATH CHALLENGE are placed in a pot. If one letter is selected at random, what is the probability that it is an A or an L? Express your answer as a common fraction.

Answer: $\frac{5}{17}$

Solution: There are 3 A's and 2 L's which mean that there are 5 letters that is either an A or an L.

The probability is: $\frac{5}{17}$ where there are 17 total letters and 5 desirable letters.

3. The four side lengths of a square are $(6x - 2)$, $(3x + 10)$, $(4x + 6)$, and $(7x - 6)$. What is the perimeter of the square? Express your answer as a positive integer.

Answer: 88

Solution: Setting any of the two side lengths equal will yield the value of x.
$6x - 2 = 3x + 10$; $3x = 12$; $x = 4$
Thus, one of the side lengths is $s = 6x - 2 = 6 \times 4 - 2 = 24 - 2 = 22$.

The perimeter of the square is $4s = 4 \times 22 = 88$.

4. The edge lengths of a rectangular solid are 3, 5 and 11. What is the positive difference between the numerical value of the volume and the numerical value of the total surface area?

Answer: 41

Solution: The volume of a rectangular solid is L × W × H = 3 × 5 × 11 = 165.
The surface area of a rectangular solid is 2(LW + WH + LH) = 2(3 × 5 + 5 × 11 + 3 × 11) = 2(15 + 55 + 33) = 2(103) = 206

The positive difference of these two values is: 206 – 165 = 41.

5. A car travels 2310 feet. It travels at a constant rate of 66 inches per second. How many minutes does it take the car to reach its destination?

Answer: 7 (minutes)

Solution: Distance = Rate × Time
Time = Distance ÷ Rate

First, convert 66 inches per second in terms of feet per minute.

$$\frac{66 \text{ inches}}{1 \text{ second}} * \frac{60 \text{ seconds}}{1 \text{ minute}} * \frac{1 \text{ foot}}{12 \text{ inches}} = \frac{66}{1} * \frac{60}{1 \text{ minute}} * \frac{1 \text{ feet}}{12} = \frac{330 \text{ feet}}{1 \text{ minute}}$$

Time = (2310 feet) ÷ (330 feet per minute) = 7 minutes

6. A sandwich consists of one kind of bread, one kind of cheese, one kind of vegetable and one kind of meat. If there are three different breads, four different cheeses, two different vegetables, and three different meats, how many unique sandwiches can one make?

Answer: 72 (sandwiches)

Solution: By the Fundamental Theorem of Counting:
There are (3 × 4 × 2 × 3) or 72 ways to make a unique sandwich.

7. The number 1001 = 7 × 11 × 13. What is the greatest prime factor of 183,183?

Answer: 61

Solution: 183,183 = 183 × 1001 = (3 × 61) × (7 × 11 × 13)
Thus, 61 is the greatest prime factor of 183,183.

8. Given the set: {20, 20, 21, 23, 25, 28, 31}, what is the sum of the set's mean, mode and range?

Answer: 55

Solution:

Mean: $\frac{20+20+21+23+25+28+31}{7} = \frac{168}{7} = 24$

Mode: 20

Range: 31 – 20 = 11

Mean + Mode + Range = 24 + 20 + 11 = 55.

9. Find the value of $\frac{1+2+3+...+98+99+100}{101}$.

Answer: 50

Solution: The numerator of the fraction is the sum of an Arithmetic Sequence:

$$\frac{100(1+100)}{2} = \frac{100*101}{2} = 50*101 = 5050$$

Therefore, the answer is $\frac{5050}{101} = 50$.

10. How many ways can the letters in the word ARRAY be arranged?

Answer: 30 (ways)

Solution: There are 5 total letters, thus, there are 5! ways to arrange them. There are 2 A's and 2 R's which can both be rearranged 2! Ways.

Thus, the total number of ways is $\frac{5!}{2!*2!} = \frac{120}{4} = 30$

2013 Countdown Round Solutions

1. 98
2. 50%
3. 4 (factors)
4. 30
5. 9π
6. $4.00
7. 100
8. 100 (jellybeans)
9. 6 (feet^2)
10. 40%
11. $4.50
12. 1800 (seconds)
13. 126 (miles)
14. 55
15. 64
16. 60°
17. 27
18. 125 (inches)
19. 29
20. 25
21. 4:21 PM
22. 3.14
23. 100
24. 3 (numbers)
25. 4
26. 24 (combinations)
27. 343 (inches^3)
28. 56 (ounces)
29. 720°
30. 3
31. 65
32. 1760 (yards)
33. 480
34. 13 (days)
35. ¼
36. $\frac{8}{15}$
37. $1350.00
38. 11 (marbles)
39. 24 (inches^2)
40. 7
41. 18
42. 391
43. 3
44. 11:11 AM
45. 6π
46. 1,111
47. 102
48. 3
49. 15 (minutes)
50. $65.00
51. 16
52. 3:10 AM
53. 66
54. 9
55. 3 (hours)
56. 4
57. 180°
58. 4
59. 1
60. 345 (students)
61. 10
62. 0.15
63. 63 (inches^2)
64. 7 (yards)
65. 15
66. 6:32 PM
67. $\frac{5}{14}$
68. 30
69. 4,000,169
70. 28 (apples)
71. 19
72. $\frac{1}{12}$
73. 8640
74. 13
75. 46%
76. 0
77. 8
78. 0
79. 360°
80. 64

1. What is the largest two-digit number that is divisible by 7?

Answer: 98
Solution: $100 \div 7 = 14$ R 2. $7 \times 14 = 98$

2. What percent of 10 is 5?

Answer: 50%
Solution: $\frac{5}{10} * 100 = 50$

3. How many factors does the number 6 have?

Answer: 4 (factors)
Solution: $6 = 2^1 3^1$. Number of Factors = $(1 + 1) \times (1 + 1) = 2 \times 2 = 4$

4. What is the product of the first three prime numbers?

Answer: 30
Solution: 2, 3, and 5 are the first three prime numbers. $2 \times 3 \times 5 = 30$.

5. What is the area of a circle with radius 3? Express your answer in terms of π.

Answer: 9π
Solution: Area of a circle = $\pi r^2 = \pi \times 3^2 = 9\pi$.

6. John had $5.00 when he entered a toy store. He spent 20% of his money on a toy truck and had exactly enough money to buy a toy car. How much did the toy car cost in dollars?

Answer: $4.00
Solution: Toy Truck = $5.00 × 20% = $1.00
Toy Car = $5.00 – $1.00 = $4.00

7. Find the value of $7^2 + 3 \times 17$.

Answer: 100
Solution: $7^2 + 3 \times 17 = 49 + 51 = 100$.

8. Happy was so happy that he ate 25 jellybeans out of a jar. His sister, who was also happy, ate 50 jellybeans out of the jar. The rest, which is 25% of the original amount of jellybeans, was eaten by their mom. How many jellybeans were there in the jar originally?

Answer: 100 (jellybeans)
Solution: 25 + 50 = 75. This is (100 - 25)% or 75% of the total jellybeans. Thus, there are 100 original jellybeans.

9. A right triangle has legs 3 feet and 4 feet, and hypotenuse 5 feet. What is the area of the triangle in square feet?

Answer: 6 (feet^2)

Solution: $$\text{Area of a right triangle} = \frac{l_1 l_2}{2} = \frac{3 * 4}{2} = 6$$

10. If Mary can hold 6 cups in one hand, what percent of a box with 30 cups can she hold in two hands?

Answer: 40%

Solution: $$\frac{6 * 2}{30} * 100 = 40$$

11. If one newspaper and two books cost $7.00 while two newspapers and one book cost $6.50, how much do one newspaper and one book cost in dollars?

Answer: $4.50
Solution: Adding the two equations, three newspapers and three books cost ($7.00 + $6.50) or $13.50.

1 N + 2 B = $7.00
+ (2 N + 1 B = $6.50)
3 N + 3 B = $13.50

Thus, dividing this by three yields: one newspaper and one book cost $4.50.

12. How many seconds are there in ½ hour?

Answer: 1800 (seconds)
Solution: ½ hour = (½ × 60) minutes = 30 minutes = (30 × 60) seconds = 1800 seconds

13. The cross country team ran 4.5 miles each day in February of 2013. How many miles did they run total in February 2013?

Answer: 126 (miles)
Solution: 4.5 miles per day × 28 days = 126 miles

14. Find the value of 1 + 2 + 3 + … + 8 + 9 + 10.

Answer: 55

Solution: Sum of arithmetic sequence (See Formula and Tips): $\frac{10(1+10)}{2} = 55$

15. Find the perimeter of a regular octagon with side length 8.

Answer: 64
Solution: s = 8. p = 8s = 8 × 8 = 64

16. What is the average of the angles in a triangle with angles: 80°, 75° and 25°?

Answer: 60°
Solution: The sum of the angles in any triangle is 180°.
Thus, the average is 180° ÷ 3 = 60°.

17. What is the product of the sum and the positive difference of the numbers 3 and 6?

Answer: 27
Solution: Sum = 3 + 6 = 9. Positive Difference = 6 – 3 = 3.
Product of the Sum and the Positive Difference = 9 × 3 = 27

18. If a tree is 2 yards, 4 feet and 5 inches tall, how tall is the tree in inches?

Answer: 125 (inches)
Solution: 2 yards = (2 × 3) feet = 6 feet = (6 × 12) inches = 72 inches
4 feet = (4 × 12) inches = 48 inches
72 + 48 + 5 = 125 inches

19. Find the positive difference of the two numbers 201.3 and 230.3.

Answer: 29

Solution: 230.3 – 201.3 = 29

20. What is the area of a square with perimeter 20?

Answer: 25
Solution: P = 4s = 20; s = 20 ÷ 4 = 5; $A = s^2 = 5^2 = 25$.

21. If Cindy starts eating lunch at 3:43 PM and she eats for 38 minutes, at what time does she finish eating?

Answer: 4:21 PM
Solution: 3:43 PM + 38 minutes = 3:81 = 4:21 PM
Note: 1 hour = 60 minutes

22. Round π to the nearest hundredth.

Answer: 3.14
Solution: π = 3.141… Since the thousandth place is 1, the number rounds down to 3.14.

23. Find the value of 4 × 4 × 4 + 6 × 6.

Answer: 100
Solution: 4 × 4 × 4 + 6 × 6 = (4 × 4 × 4) + (6 × 6) = 64 + 36 = 100.

24. If Zach eats between 6 and 10 candies, how many different numbers of candies could he have eaten?

Answer: 3 (numbers)
Solution: (10 – 6) – 1 = 4 – 1 = 3
Or count: 7, 8, 9 = 3 numbers

25. What is the Greatest Common Factor (GCF) of 12, 16 and 20?

Answer: 4
Solution: $12 = 2^2 3^1$. $16 = 2^4$. $20 = 2^2 5^1$. The three numbers share $2^2 = 4$.
The GCF is 4.

26. Carlos wears one shirt and one pair of pants every day. If he has 8 shirts and 3 pairs of pants, how many different combinations of clothes can he wear?

Answer: 24 (combinations)
Solution: By the Fundamental Theorem of Counting, $8 \times 3 = 24$ combinations

27. What is the volume of a cube with side length 7 inches? Express your answer in cubic inches.

Answer: 343 (inches^3)
Solution: $V = s^3 = 7^3 = 343$.

28. How many ounces are in 3.5 pounds?

Answer: 56 (ounces)
Solution: 1 pound = 16 ounces. 3.5 pounds = (3.5×16) ounces = 56 ounces

29. What is the sum of the interior angles in a regular hexagon?

Answer: 720°
Solution: Total Degrees = $(n - 2)180 = (6 - 2)180 = 4 \times 180 = 720°$

30. What is the remainder when 1,234,567,893 is divided by 5?

Answer: 3
Solution: The last digit is 3 so the remainder is the same as when 3 is divided by 5.
$3 \div 5 = 0$ R 3

31. What is the Least Common Multiple (LCM) of 5 and 13?

Answer: 65
Solution: 5 and 13 are prime numbers. Thus, the LCM is the product of 5 and 13 or
$5 \times 13 = 65$.

32. How many yards are in a mile?

Answer: 1760 (yards)
Solution: 1 yard = 3 feet. 1 mile = 5280 feet.
1 mile = 5280 feet = $(5280 \div 3)$ yards = 1760 yards

33. Find the volume of a rectangular solid with lengths 6, 8 and 10.

Answer: 480
Solution: V = $lwh = 6 \times 8 \times 10 = 480$.

34. How many days are in 312 hours?

Answer: 13 (days)
Solution: 1 day = 24 hours. 312 hours = (312 ÷ 24) days = 13 days

35. When two coins are flipped, what is the probability that they are both heads? Express your answer as a common fraction.

Answer: $\frac{1}{4}$

Solution: $$\text{P}(2 \text{ heads}) = \text{P}(\text{head}) * \text{P}(\text{head}) = \frac{1}{2} * \frac{1}{2} = \frac{1}{4}$$

36. A bag of gummy worms contains 3 yellow, 4 blue and 8 red worms. What is the probability of drawing a red worm? Express your answer as a common fraction.

Answer: $\frac{8}{15}$
Solution: There are (3 + 4 + 8) or 15 total gummy worms.

Thus, the probability of drawing a red worm is $\frac{8}{15}$.

37. An Algebra book costs $45 each. Each box contains 6 Algebra books. What is the total cost of the books if there are 5 boxes? Express your answer in dollars.

Answer: $1350.00
Solution: 5 boxes = (5 × 6) books = 30 books = (30 × 45) = $1350.00

38. A bag has 3 red marbles, 6 blue marbles and 4 green marbles. What is the least number of marbles that James needs to take out in order to ensure that he has at least 1 red marble?

Answer: 11 (marbles)
Solution: In order to guarantee that he has at least 1 red marble, he must take out one more than the number of marbles that are not red. Blue + Green + 1 = 6 + 4 + 1 = 11 marbles.

39. Buy-A-Cube-Box sells boxes that are cubes. If the side length of a box is 2 inches, what is the surface area of this closed box? Express your answer in square inches.

Answer: 24 (inches^2)
Solution: Surface Area of a Cube = $6s^2 = 6 \times 2^2 = 24$ inches^2

40. Solve for q: $7q - 31 = 18$.

Answer: 7
Solution: $7q - 31 = 18$. Adding 31 to both sides: $7q = 49$.
Dividing both sides by 7: $q = 7$.

41. What is the sum of all of the factors of 10?

Answer: 18
Solution: $10 = 2^1 5^1$. Sum of factors: $(2^0 + 2^1) \times (5^0 + 5^1) = 3 \times 6 = 18$.

42. Find the value of 17×23.

Answer: 391
Solution: Difference of Two Squares: $a^2 - b^2 = (a - b)(a + b) = (20 - 3) \times (20 + 3) = 20^2 - 3^2 = 400 - 9 = 391$.

43. If $n = 4$, what is the value of $n^2 - 5n + 7$?

Answer: 3
Solution: $n^2 - 5n + 7 = 4^2 - (5 \times 4) + 7 = 16 - 20 + 7 = 3$.

44. Joe takes 26 minutes to drive from home to work. If work starts at 11:37 AM, what time should Joe leave his house in order to arrive at work on time?

Answer: 11:11 AM
Solution: He should leave 26 minutes before 11:37 AM: 11:37 AM – 26 minutes = 11:11 AM.

45. What is the circumference of a circle with area 9π? Express your answer in terms of π.

Answer: 6π

Solution: $A = \pi r^2 = 9\pi$. $r^2 = 9$. $r = 3$.
$C = 2\pi r = 2\pi \times 3 = 6\pi$.

46. Find the value of 12,590 – 11,480 + 1.

Answer: 1,111
Solution: 12,590 – 11,480 + 1 = 1,110 + 1 = 1,111

47. What is the range of the set: {31, 22, 58, 99, 79, 124}?

Answer: 102
Solution: {31, 22, 58, 99, 79, 124} = {22, 31, 58, 79, 99, 124}
Range = 124 – 22 = 102

48. Solve for x: $13x - 3 = 36$.

Answer: 3
Solution: Adding 3 to both sides: $13x = 39$. Divide both sides by 13: $x = 3$.

49. It takes Celine one hour to travel 60 miles. How many minutes does it take Celine to travel 15 miles?

Answer: 15 (minutes)
Solution: It takes her 1 hour (or 60 minutes) to travel 60 miles. This means it takes her 1minute to travel 1 mile. Thus, it takes her 15 minutes to travel 15 miles.

50. A suit sells for \$100. If the manager decides to give a discount of 35% off on the suit, how much would the suit sell for now? Express your answer in dollars.

Answer: \$65.00
Solution: If the discount is 35% off, the new price would be 65% of the original price.
\$100 × 65% = \$65.00

51. Find the area of a square with perimeter 16.

Answer: 16
Solution: $P = 4s = 16$. $s = 4$.
$A = s^2 = 4^2 = 16$.

52. A movie starts at 11:47 PM and it runs for 203 minutes. At what time does the movie end?

Answer: 3:10 AM
Solution: 203 minutes = 3 hours and 23 minutes. (Note: 1 hour = 60 minutes)
23 minutes after 11:47 PM is 12:10 AM. 3 hours after 12:10 AM is 3:10 AM.

53. What is six more than five times a dozen?

Answer: 66
Solution: $6 + (5 \times 12) = 6 + 60 = 66$

54. What is 15% of 60?

Answer: 9
Solution: $\frac{15}{100} * 60 = 9$

55. Jarvis can drive 5 miles in 20 minutes. How many hours does it take him to drive 45 miles?

Answer: 3 (hours)
Solution: $45 \text{ miles} * \frac{20 \text{ minutes}}{5 \text{ miles}} * \frac{1 \text{ hour}}{60 \text{ minutes}} = 3 \text{ hours}$

56. Find the value of $[(2 \times 12) - 22]^2$.

Answer: 4
Solution: $[(2 \times 12) - 22]^2 = [24 - 22]^2 = 2^2 = 4$

57. What is the sum of the interior angles in a triangle?

Answer: 180°
Solution: $(n - 2)180 = (3 - 2)180 = 180°$

58. If $A + B + C = 10$ and $A = C = 3$, what is the value of B?

Answer: 4

Solution: A + B + C = 10
3 + B + 3 = 10
6 + B = 10.
Subtracting 6 from both sides: B = 10 – 6 = 4.

59. What is the remainder when 2013 × 2015 is divided by 2?

Answer: 1
Solution: 2013 and 2015 are both odd numbers.
The product of any two odd numbers will be odd (See Formula and Tips).
Therefore, the product will have a remainder of 1 when divided by 2.

60. If there are 15 buses and each bus can hold 23 students, what is the maximum number of students that the buses can hold?

Answer: 345 (students)
Solution: 15 × 23 = 345 students

61. What is the remainder when 7^2 is divided by 13?

Answer: 10
Solution: $7^2 = 49$. 49 ÷ 13 = 3 R 10

62. Round $\frac{5}{33}$ to the nearest hundredth.

Answer: 0.15
Solution: $\frac{5}{33} * \frac{3}{3} = \frac{15}{99}$
Thus, $\frac{5}{33}$ = 0.151515… This is rounded to 0.15 for the nearest hundredth.

63. What is the area of a triangle with base 7 inches and height 18 inches? Express your answer in square inches.

Answer: 63 (inches^2)
Solution: $A = \frac{bh}{2} = \frac{7 * 18}{2} = 63$

64. How many yards are in 252 inches?

Answer: 7 (yards)
Solution: Note: 1 foot = 12 inches. 1 yard = 3 feet.
252 inches = (252 ÷ 12) feet = 21 feet = (21 ÷ 3) yards = 7 yards

65. What is the mode in the set {11, 12, 14, 15, 18, 19, 15, 20}?

Answer: 15
Solution: 15 appears twice while the other numbers only appear once. Thus, 15 is the mode.

66. Robert goes to see a 169-minute movie. If the movie starts at 3:43 PM, what time does the movie end?

Answer: 6:32 PM
Solution: 169 minutes = 3 hours minus 11 minutes
3:43 PM minus 11 minutes = 3:32 PM
3:32 PM plus 3 hours = 6:32 PM

67. The letters in the words COUNTDOWN ROUND are placed in a pot. If a letter is chosen at random, what is the probability that a vowel is chosen? Express your answer as a common fraction.

Answer: $\frac{5}{14}$
Solution: There are 14 letters total. There are 5 vowels (O, U, O, O, U). Thus, the probability that a vowel is chosen is $\frac{5}{14}$.

68. What is half of 60% of 100?

Answer: 30
Solution: $\frac{1}{2} * \frac{60}{100} * 100 = 30$

69. What is the value of $2000^2 + 13^2$?

Answer: 4,000,169
Solution: $2000^2 + 13^2$ = 4,000,000 + 169 = 4,000,169

70. Johnny eats 1 apple on Monday. On Tuesday, he eats 2 apples and on Wednesday, he eats 3 apples. If he continues this pattern, how many apples will he have eaten in one week?

Answer: 28 (apples)
Solution: The answer is 1 + 2 + 3 + … + 7.
Using the sum of an arithmetic sequence, $\frac{7(1+7)}{2} = 28$

71. In the arithmetic sequence: {1, 4, 7, 10, … }, what is the value of the 7th term?

Answer: 19
Solution: The first term is 1 and the common difference is 4 – 1 = 3.
$a_n = a_1 + (n - 1)d$
Thus, the 7th term is $a_7 = 1 + (7 - 1)3 = 1 + 6 \times 3 = 1 + 18 = 19$

72. The probability that it rains on Monday is ⅕. The probability that it doesn't rain on Tuesday is $\frac{7}{12}$. What is the probability that it rains on both Monday and Tuesday? Express your answer as a common fraction.

Answer: $\frac{1}{12}$

Solution: $\text{P(Rain on Tuesday)} = 1 - \frac{7}{12} = \frac{5}{12}$
$\text{P(Rain on Both Days)} = \text{P(Rain on Monday)} * \text{P(Rain on Tuesday)} = \frac{1}{5} * \frac{5}{12} = \frac{1}{12}$

73. How many minutes are in 6 days?

Answer: 8640 (minutes)
Solution: $\frac{6 \text{ days}}{1} * \frac{24 \text{ hours}}{1 \text{ day}} * \frac{60 \text{ minutes}}{1 \text{ hour}} = 8640 \text{ minutes}$

74. Solve for *r*. $7r - 2^4 = 75$.

Answer: 13
Solution: $7r - 2^4 = 75$. $7r - 16 = 75$.
Adding 16 to both sides: $7r = 75 + 16 = 91$.
Dividing both sides by 7: $r = 13$.

75. What percent of 50 is 23?

Answer: 46%
Solution: $\frac{23}{50} * 100 = 46$

76. What is the remainder when 13^{13} is divided by 13?

Answer: 0
Solution: 13^{13} is divisible by 13. Thus, the remainder is 0 when divided by 13.

77. The area of a square is 36. If an equilateral triangle has the same perimeter as the square, what is the side length of the equilateral triangle?

Answer: 8
Solution: $A = s^2 = 36$. $s = 6$
$P_{Square} = 4s = 4 \times 6 = 24 = P_{Triangle}$
Side length of the triangle = $P_{Triangle} \div 3 = 24 \div 3 = 8$.

78. Find the digit n that satisfies the following equation: $1{,}234{,}56\underline{n} = 123{,}456 \times 10$

Answer: 0
Solution: $1{,}234{,}56\underline{n} = 123{,}456 \times 10 = 1{,}234{,}560$
Thus, the value of n is 0.

79. What is the sum of the angles in a regular quadrilateral?

Answer: 360°
Solution: $(n - 2)180 = (4 - 2)180 = 2 \times 180 = 360°$

80. Find the value of $1 + 3 + 5 + \ldots + 11 + 13 + 15$.

Answer: 64
Solution: The sum of the first eight odd numbers is $n^2 = 8^2 = 64$.

2014 Answer Keys and Solutions

2014 Individual Round Solutions

Answer Key

1. d) 14
2. b) 7
3. c) 21
4. e) 81 units^2
5. b) 343
6. a) 132
7. e) 9:05 PM
8. a) $\frac{5}{26}$
9. c) \$14
10. e) 1680 seconds
11. c) 216
12. b) 8
13. e) 20
14. a) ⅓
15. c) 640 fl. oz.
16. d) 256π
17. d) 75,735
18. b) 192
19. b) 3
20. c) 36 boys
21. 14 (diagonals)
22. 20 (times)
23. 1760
24. 2:30 PM
25. 10 (committees)
26. ½
27. 294 (meters^2)
28. 806
29. 4
30. 84 (packs)

1. Solve: $6 + 2 \times (3^2 - 5)$.

Answer: d) 14

Solution: Using the Order of Operations:
$6 + 2 \times (3^2 - 5) = 6 + 2 \times (9 - 5) = 6 + 2 \times 4 = 6 + 8 = 14$

2. How many prime numbers are between 30 and 60?

Answer: b) 7

Solution: The prime numbers between 30 and 60 are: 31, 37, 41, 43, 47, 53 and 59. There are 7 prime numbers.

3. Given a set: {17, 33, 5, 25}. Find the median.

Answer: c) 21

Solution: Since there is an even number of terms in the set, the median is the average of the two middle numbers. Rearranging {17, 33, 5, 25}, gives {5, 17, 25, 33}.
Thus, $\text{Median} = \frac{17+25}{2} = \frac{42}{2} = 21$

4. What is the area of a square with perimeter 36 units?

Answer: e) 81 units^2

Solution: P = 4s = 36; s = 36 ÷ 4 = 9 units
$A = s^2 = 9^2 = 9 \times 9 = 81$ units^2

5. 485 pennies and 123 dimes have the same value as *n* nickels. What is the value of *n*?

Answer: b) 343

Solution: 485 pennies (each worth 1 cent) is equivalent to 485 cents. 123 dimes (each worth 10 cents) is equivalent to 1230 cents. Thus, the total amount is 485 + 1230 = 1715 cents.
A nickel is worth 5 cents. Therefore, 1715 ÷ 5 = 343 nickels.

6. Find the perimeter of the triangle shown below. Note: Figure not drawn to scale.

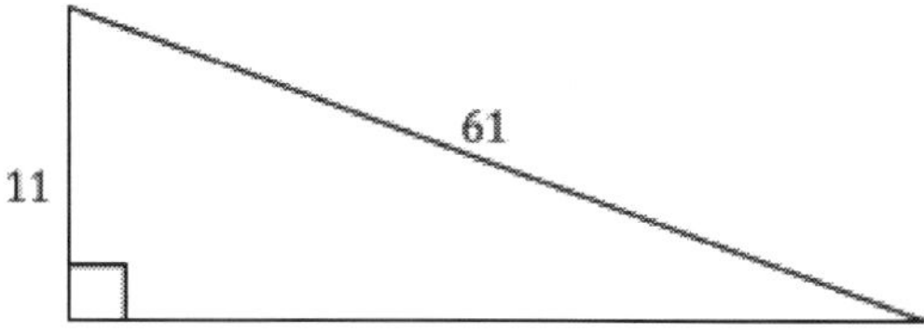

Answer: a) 132

Solution: The pythagorean triple that matches this right triangle is 11-60-61 (See Formula and Tips). The other leg can also be found using the pythagorean theorem and the difference of two squares:

$$\sqrt{61^2 - 11^2} = \sqrt{(61-11)(61+11)} = \sqrt{50 * 72} = \sqrt{25 * 144} = 5 * 12 = 60$$

Thus, the perimeter of the right triangle is (11 + 60 + 61) = 132.

7. Maui Math Cinemas is showing a movie that is 82 minutes in length. If the movie starts at 7:43 PM, what time will the movie finish?

Answer: e) 9:05 PM

Solution: 82 minutes is equivalent to 1 hour and 22 minutes.
22 minutes after 7:43 PM is 8:05 PM. 1 hour after 8:05 PM is 9:05 PM.

8. All the letters in the alphabet are placed in a bag and shuffled. If John reaches in the bag and draws one letter, what is the probability that he will draw a vowel? Assume that the letter "y" is not a vowel.

Answer: a) $\frac{5}{26}$

Solution: There are 26 letters in the alphabet and 5 of those 26 letters are vowels (A, E, I, O, U). Thus, the probability that John draws a vowel is $\frac{5}{26}$.

9. Josh has $10 more than a third of Jamie's money. If Jamie has $12, how much money does Josh have?

Answer: $14

Solution: Jamie = $12

$$\text{Josh} = 10 + \frac{\text{Jamie}}{3} = 10 + \frac{12}{3} = 10 + 4 = 14$$

Josh = $14

10. Antonio takes a quick jog around a rectangular track with dimensions 300 feet by 400 feet. If he starts at one corner jogging at a speed of 50 feet per minute, how long does it take him to return back to his starting point?

Answer: e) 1680 seconds

Solution: The perimeter of a rectangle is P = $2l + 2w$. If $l = 400$ and $w = 300$, the perimeter is P = 2 × 400 + 2 × 300 = 800 + 600 = 1400 feet.
If Distance = 1400 feet and Rate = 50 feet per minute, Time can be found using Distance = Rate × Time.

$$\text{Time} = \frac{\text{Distance}}{\text{Rate}} = \frac{1400}{50} = 28 \text{ minutes}$$

Since 28 minutes is not one of the answers, convert it to seconds.
28 minutes = (28 × 60) seconds = 1680 seconds.

11. Find the product of the Least Common Multiple (LCM) and the Greatest Common Factor (GCF) of 12 and 18.

Answer: c) 216

Solution: The product of the LCM and the GCF of two numbers is simply the product of the two numbers (See Formula and Tips).
Thus, LCM × GCF = 12 × 18 = 216.
(Note: LCM = 36 and GCF = 6)

12. Solve. $\dfrac{5^3 - 7 * (3+8)}{10 - 2 * \sqrt{3+1}}$

Answer: b) 8

Solution: Using the Order of Operations:

$$\frac{5^3 - 7 * (3+8)}{10 - 2 * \sqrt{3+1}} = \frac{5^3 - 7 * 11}{10 - 2 * \sqrt{4}} = \frac{125 - 77}{10 - 2 * 2} = \frac{48}{10 - 4} = \frac{48}{6} = 8$$

13. Manny's daily outfit consists of one shirt and one pair of pants. If he has four different shirts and five different pairs of pants, how many unique outfits can he make?

Answer: e) 20

Solution: Using the Fundamental Theorem of Counting,
the number of unique outfits is $4 \times 5 = 20$.

14. A regular die is rolled. What is the probability that the number rolled is a multiple of three?

Answer: a) ⅓

Solution: The number of possibilities is 6 (1, 2, 3, 4, 5, 6). There are 2 multiples of three (3 and 6).

Thus, the probability that the number rolled is a multiple of 3 is $\frac{2}{6} = \frac{1}{3}$.

15. How many fluid ounces are in five gallons?

Answer: c) 640 fl. oz.

Solution: Using the conversion factors: 1 gallon = 4 quarts, 1 quart = 2 pints, 1 pint = 2 cups, 1 cup = 8 fl. Oz.

$$5 \text{ gallons} * \frac{4 \text{ quarts}}{1 \text{ gallon}} * \frac{2 \text{ pints}}{1 \text{ quart}} * \frac{2 \text{ cups}}{1 \text{ pint}} * \frac{8 \text{ fl. oz.}}{1 \text{ cup}} = (5*4*2*2*8) \text{ fl. oz.} = 640 \text{ fl. oz.}$$

16. A circular wheel has circumference 16π. Mary wants to buy four of these wheels for her car. What is the total area of all the wheels that she wants to purchase?

Answer: d) 256π

Solution: $C = 2\pi r = 16\pi$. Dividing both sides by 2π: $r = 8$.
The area of one wheel is $A = \pi r^2 = \pi \times 8^2 = 64\pi$.
Thus, the area of four wheels is $4A = 4 \times 64\pi = 256\pi$

17. Which of the following numbers is divisible by 9?

Answer: d) 75,735

Solution: The divisibility rule for 9 is if the sum of the digits of the number is divisible by 9.
75,735 is divisible by 9 because (7 + 5 + 7 + 3 + 5) = 27 and 27 is divisible by 9.

18. Find the area of the given rectangle shown below. Note: Figure not drawn to scale.

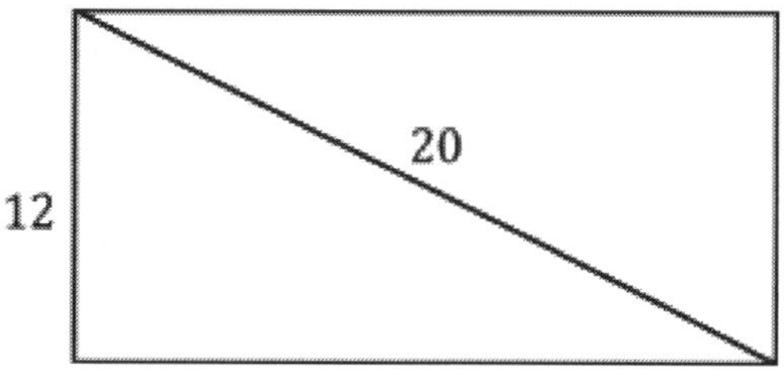

Answer: b) 192

Solution: The length and the width of the rectangle along with the diagonal form a right triangle. Thus, the missing side length of the rectangle can be found using the pythagorean theorem.
The pythagorean triple that matches this right triangle is 12-16-20 which is 3-4-5 multiplied by a factor of 4 (See Formula and Tips). The other leg can also be found using the pythagorean theorem and the difference of two squares:
$\sqrt{20^2 - 12^2} = \sqrt{(20-12)(20+12)} = \sqrt{8 * 32} = 16$
Thus, the area of the rectangle is A = lw = 12 × 16 = 192.

19. Given a set: {23, 7, 29, 28, 13}. Find the positive difference between the mean and the median.

Answer: b) 3

Solution: Rearranging {23, 7, 29, 28, 13} gives {7, 13, 23, 28, 29}. Thus, the median is 23.
The mean is $\frac{\text{Sum of Terms}}{\text{Number of Terms}} = \frac{7+13+23+28+29}{5} = \frac{100}{5} = 20$.
Hence, the positive difference is 23 – 20 = 3.

20. The ratio of boys to girls at an elementary school is 4:5. If there are 45 girls in the school, how many boys are there?

Answer: c) 36 boys

Solution: If there are 4 boys for every 5 girls and there are 45 girls, this ratio is multiplied by a factor of (45 ÷ 5) or 9. Thus, this factor of 9 can be multiplied by 4 to give the number of boys: 4 × 9 = 36 boys.

21. How many diagonals does a heptagon have?

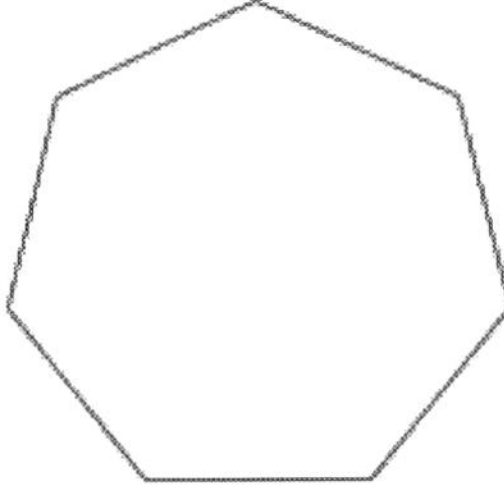

Answer: 14 (diagonals)

Solution: The formula for the number of diagonal in a convex *n*-gon is $\frac{n(n-3)}{2}$ (See Formula and Tips).

Thus, there are $\frac{7(7-3)}{2} = \frac{7 * 4}{2} = 14$ diagonals.

22. Samantha loves the number 5 so much that she wants to count how many times the digit 5 appears from 1 to 100. How many times does the digit 5 appear from 1 to 100?

Answer: 20 (times)

Solution: The digit 5 appears in the tens and the units digit.
For the units digit, it appears 10 times: 5, 15, 25, … , 95
For the tens digit, it appears 10 times: 50, 51, 52, ... , 59
Thus, it appears (10 + 10) = 20 times.

23. Rebecca has to walk from home to school. The journey is a third of a mile. To keep herself busy, she calculates that she must travel *n* feet in order to go from home to school. What is the value of *n*?

Answer: 1760

Solution: 1 mile = 5280 feet.
Since she traveled ⅓ of a mile, she traveled 5280 ÷ 3 = 1760 feet.

24. Carl drives at a constant rate of 50 miles per hour from point A to point B. The distance he must travel is 125 miles. If he leaves point A at 12:00 PM, what time will he arrive at point B?

Answer: 2:30 PM

Solution: The distance is 125 miles and the rate is 50 miles per hour. Thus, the time can be found using Distance = Rate × Time.
Time = $\frac{125}{50} = \frac{5}{2}$ hours. This is equivalent to 2 hours and 30 minutes.
2 hours and 30 minutes after 12:00 PM is 2:30 PM.

25. A classroom has 5 students: Al, Bob, Carrie, Don and Effie. How many different 2-person committees can be chosen from these 5 students? Note: A committee of Al-Bob is the same as a committee of Bob-Al.

Answer: 10 (committees)

Solution: For each student, there are (5 - 1) or 4 other people that can join to form a 2-person committee. Thus, 5 × 4 = 20. However, this number is divided by 2 because a committee of Al-Bob is the same as a committee of Bob-Al.
Therefore, 20 ÷ 2 = 10 committees.

26. Two fair coins are flipped. What is the probability that one head and one tail are flipped. Express your answer as a common fraction.

Answer: ½

Solution: There are $2^2 = 4$ possibilities for the 2 coin flips. HT and TH are the only two possibilities that have one head and one tail.
Therefore, the probability is $\frac{2}{4} = \frac{1}{2}$.

27. What is the surface area of a cube with volume 343 cubic meters? Express your answer in square meters.

Answer: 294 meters^2

Solution: $V = s^3 = 343$. Knowing that $7^3 = 343$, $s = 7$.
Thus, Surface Area = $6s^2 = 6 \times 7^2 = 6 \times 49 = 294$ meters^2.

28. Find the sum: 11 + 12 + 13 + … + 39 + 40 + 41.

Answer: 806

Solution: There are (41 – 11) + 1 = 31 numbers.
$n = 31$, $a_1 = 11$ and $a_{31} = 41$.
The formula for the sum of an arithmetic sequence is

$$S_n = \frac{n(a_1 + a_n)}{2} = \frac{31 * (11 + 41)}{2} = \frac{31 * 52}{2} = 31 * 26 = 806$$

29. What is the units digit of the decimal representation of 4^{11} ?

Answer: 4

Solution: $4^1 = 4$, ends in 4. $4^2 = 16$, ends in 6. $4^3 = 64$, ends in 4.
Thus, the pattern is that 4^n ends in 4 when *n* is odd and 4^n ends in 6 when *n* is even.
Therefore, 4^{11} ends in or has a units digit of 4.

30. Pencils are sold in packs of twelve. If Sally needs to buy 1,000 pencils, what is the least number of packs of pencils that she must buy?

Answer: 84 (packs)

Solution: 1000 ÷ 12 = 83 R 4
Thus, she must buy 84 packs.

2014 Team Round Solutions

Answer Key

1. 72 (animals)

2. 12

3. 700

4. 180 (ways)

5. 1

6. 10:29:10 AM

7. 38

8. 65

9. 5 (chickens)

10. 12 (outfits)

1. There are cows, chickens and pigs in a barn. The ratio of cows to chickens to pigs is 5:17:14. The number of chickens in the barn is 34. How many total animals are there in the barn?

Answer: 72 (animals)

Solution: Cows:Chickens:Pigs = 5:17:14 and Chickens = 34. Thus, the factor multiplied to the ratio is 34 ÷ 17 = 2.
This factor can be multiplied by the ratio: (5:17:14) × 2 = 10:34:28 = Cows:Chickens:Pigs
The total number of animals is 10 + 34 + 28 = 72.

2. Find the positive difference between the sum of the number of edges, the number of vertices and the number of faces of a cube, and the sum of the number of edges, the number of vertices and the number of faces of a tetrahedron.

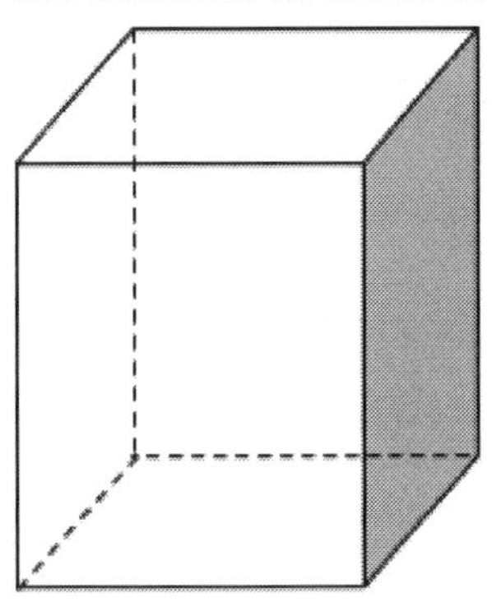

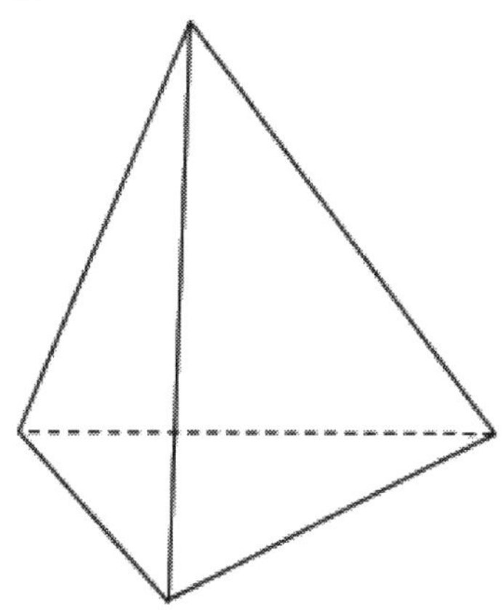

Answer: 12

Solution: Using the figures above, the following can be found:
$\text{Edges}_{\text{Cube}} = 12$, $\text{Vertices}_{\text{Cube}} = 8$, $\text{Faces}_{\text{Cube}} = 6$
$\text{Edges}_{\text{Tetrahedron}} = 6$, $\text{Vertices}_{\text{Tetrahedron}} = 4$, $\text{Faces}_{\text{Tetrahedron}} = 4$

$\text{Edges}_{\text{Cube}} + \text{Vertices}_{\text{Cube}} + \text{Faces}_{\text{Cube}} = 12 + 8 + 6 = 26$
$\text{Edges}_{\text{Tetrahedron}} + \text{Vertices}_{\text{Tetrahedron}} + \text{Faces}_{\text{Tetrahedron}} = 6 + 4 + 4 = 14$

The positive difference between these two sums is 26 – 14 = 12.

3. Find the product of the mean, median and range of the set:
{2, 0, 2, 5, 5, 20, 11, 3, 3, 19, 7, 1, 13}.

Answer: 700

Solution: Rearranging the 13 numbers: {0, 1, 2, 2, 3, 3, 5, 5, 7, 11, 13, 19, 20}

Mean = $\frac{0+1+2+2+3+3+5+7+11+13+19+20}{13} = \frac{91}{13} = 7$

Median = 5

Range = 20 – 0 = 20

Mean × Median × Range = 7 × 5 × 20 = 700

4. How many ways can the letters in the word "PEOPLE" be arranged?

Answer: 180 (ways)

Solution: There are 6! ways to arrange 6 letters. However, there are 2 P's and 2 E's that are identical when arranged. Thus, the number of ways to arrange the letters in "PEOPLE" is $\frac{6!}{2! * 2!} = \frac{720}{4} = 180$

5. What is the remainder when 9^{12} is divided by 10?

Answer: 1

Solution: When a number is divided by 10, the remainder is the units digit.

$9^1 = 9$ ends in 9. $9^2 = 81$ ends in 1. $9^3 = 729$ ends in 9.

Thus, 9^n ends in 9 when n is odd and 9^n ends in 1 when n is even.

Therefore, 9^{12} ends in 1. Hence, the remainder when 9^{12} is divided by 10 is 1.

6. A subway train in Washington D.C. departs every 2014 seconds. If a train just left at 9:55:36 AM, what time will the next subway train depart? Express your answer in the form AB:CD:EF where AB denotes the hour, CD denotes the minute and EF denotes the second.

Answer: 10:29:10 AM

Solution: 2014 seconds is equivalent to 33 minutes and 34 seconds (2014 ÷ 60 = 33 R 34).

Thus, the next train will depart 33 minutes and 34 seconds after 9:55:36 AM.

34 seconds after 9:55:36 AM is 9:56:10 AM.

33 minutes after 9:56:10 AM is 10:29:10 AM.

7. Find the perimeter of the following figure if all of the angles are either 90 or 270 degrees. Note: Figure not drawn to scale.

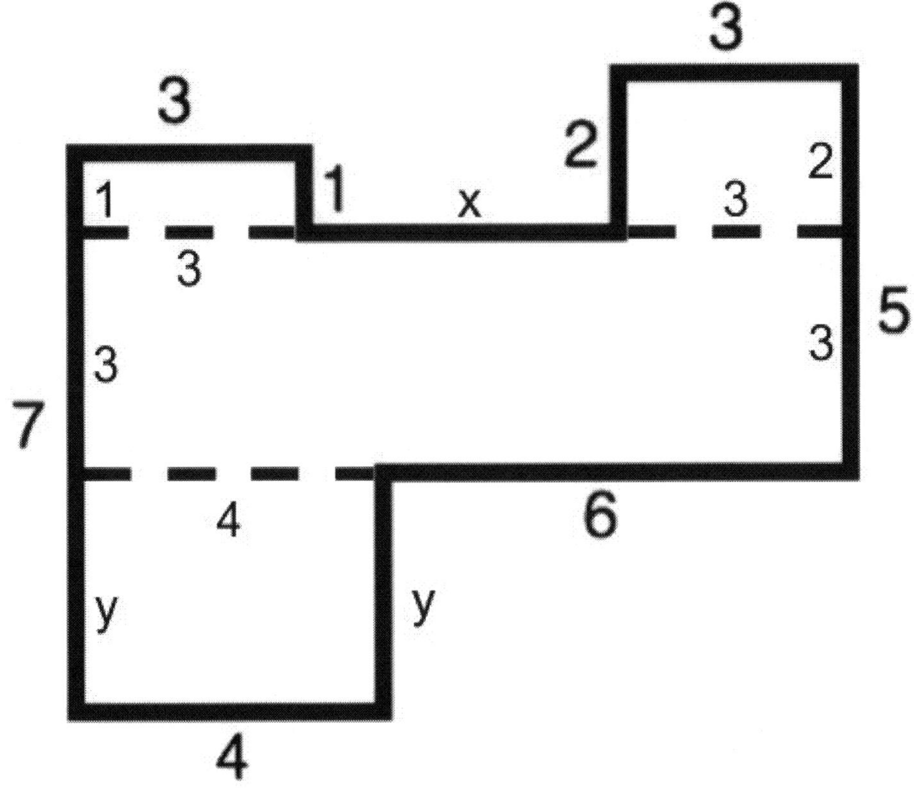

Answer: 38

Solution: The figure is only missing 2 lengths (x, y). Drawing the above lines, several rectangles are formed which means that opposite sides are equivalent.
Looking at the side with length 7, three segments make up that side: 1, 3 and y.
Thus, $1 + 3 + y = 7$; $4 + y = 7$; $y = 7 - 4 = 3$.
Looking at the horizontal line that contains 3, x, and 3, it has the same total length as the line below it that contains 4 and 6.
Thus, $3 + x + 3 = 4 + 6$; $6 + x = 10$; $x = 10 - 6 = 4$.

The missing lengths have been found and now, the perimeter can be found:
Starting from the side with length 7 and moving clockwise around the figure:
$7 + 3 + 1 + 4 + 2 + 3 + 5 + 6 + 3 + 4 = 38$.

8. Assume that each letter in the alphabet is worth its place in the alphabetical order. For example, A = 1, B = 2, …, Y = 25 and Z = 26. If this is true, what is the sum of the values of the letters in the word ALPHABET?

Answer: 65

Solution: ALPHABET contains 2 A's, 1 B, 1 E, 1 H, 1 L, 1 P and 1 T.
Going through the alphabet, A = 1, B = 2, E = 5, H = 8, L = 12, P = 16, T = 20.
A + A + B + E + H + L + P + T = 1 + 1 + 2 + 5 + 8 + 12 + 16 + 20 = 65.

9. Farmer Kai has only chickens and horses on his farm. In total, the animals have 18 heads and 62 legs. Assuming chickens have 2 legs and horses have 4 legs, how many chickens does Kai have?

Answer: 5 (chickens)

Solution: Assume that every animal on the farm is a chicken. That means there are $18 \times 2 = 36$ legs total.
However, this is $62 - 36 = 26$ legs less than the actual number of legs. A horse has $4 - 2 = 2$ more legs than a chicken. That means there are $26 \div 2 = 13$ horses on Kai's farm.
Thus, there are $18 - 13 = 5$ chickens.

10. Danny has 2 different shirts, 2 different pair of pants and 3 different hats. If an outfit consists of one shirt, one pair of pants and one hat, how many different outfits could Danny wear?

Answer: 12 (outfits)

Solution: Using the Fundamental Theorem of Counting,
Shirts × Pants × Hats = $2 \times 2 \times 3 = 12$ different outfits

2014 Countdown Round Solutions

1. ⅙
2. 9 (units)
3. 64
4. 12 (quarts)
5. 0
6. 90
7. 17 (quarters)
8. 28 (units)
9. 2:32 PM
10. 17
11. 30
12. 21
13. 139 (yards)
14. 7
15. 1
16. 4
17. –5
18. 12 (inches)
19. Friday
20. 25
21. 6 (bicycles)
22. 2 (diagonals)
23. ⅗
24. 6 (factors)
25. 6 (ways)
26. 95
27. 625
28. $\frac{9}{10}$
29. 21 (digits)
30. 64 (cents)
31. 100
32. $22,900,000
33. 2088
34. 101
35. 40
36. 30
37. 10
38. 5 (units)
39. 900 (seconds)
40. 6
41. 60°
42. 8 (days)
43. 4π
44. 1
45. 12
46. 0.375
47. 180°
48. 720
49. 13
50. 32 (ounces)
51. 5 (toy cars)
52. 4 (cars)
53. 110
54. 0
55. ½
56. 11:05 PM
57. 6
58. 51 (cents)
59. 7200 (seconds)
60. 2
61. ¼
62. 360
63. 383
64. $1600
65. 2 (diagonals)
66. 6
67. 204
68. 97 (minutes)
69. 15
70. 2 (numbers)
71. 3 (factors)
72. ⅙
73. 26,400 (feet)
74. 60%
75. 4028
76. 96
77. $14.30
78. 6 (pints)
79. 9
80. 0

2014 Countdown Round Solutions

1. A fair die is rolled. What is the probability that the number 2 is rolled? Express your answer as a common fraction.

Answer: $\frac{1}{6}$
Solution: There are 6 possibilities (1, 2, 3, 4, 5, 6) and only 1 of these possibilities is the number 2. Thus, the probability is $\frac{1}{6}$.

2. An equilateral triangle has perimeter 27 units. What is the length of one of its sides?

Answer: 9 (units)
Solution: An equilateral triangle has 3 equal sides.
Thus, p = 3s = 27 and s = 27 ÷ 3 = 9 units.

3. What is the value of $2^7 - 2^6$?

Answer: 64
Solution: $2^7 = 128$ and $2^6 = 64$
Thus, $2^7 - 2^6 = 128 - 64 = 64$.

4. For a recipe, Jacob needs 3 gallons of milk or *n* quarts of milk. What is the value of *n*?

Answer: 12 (quarts)
Solution: There are 4 quarts in 1 gallon.
Therefore, 3 gallons = (3 × 4) quarts = 12 quarts.

5. Two dice are rolled. What is the probability that the sum of the numbers rolled is 1? Express your answer in simplest form.

Answer: 0
Solution: The minimum number that can be rolled for each die is 1. Thus, the smallest sum that can be made is 2 (1 + 1 = 2). Therefore, it is impossible to roll two numbers with a sum of 1.
Thus, the probability is 0.

6. Find the mean of the set: {90, 87, 93}.

Answer: 90

Solution: Mean = $\dfrac{90+87+93}{3} = \dfrac{270}{3} = 90$

7. How many quarters are needed to exchange for 425 pennies?

Answer: 17 (quarters)
Solution: A quarter is worth 25 cents and a penny is worth 1 cent.
Thus, $425 \div 25 = 17$ quarters.

8. If a square has area 49 units^2, what is its perimeter?

Answer: 28 (units)
Solution: $A = s^2 = 49$. $s = 7$
$P = 4s = 4 \times 7 = 28$ units.

9. Bernard sees a movie that runs for 92 minutes. If he starts watching the movie at 1:00 PM, what time will the movie end?

Answer: 2:32 PM
Solution: 92 minutes is equivalent to 1 hour and 32 minutes.
1 hour and 32 minutes after 1:00 PM is 2:32 PM.

10. What is the sum of all the prime numbers between 1 and 10?

Answer: 17
Solution: The prime numbers between 1 and 10 are 2, 3, 5 and 7.
Thus, $2 + 3 + 5 + 7 = 17$.

11. What is the Least Common Multiple (LCM) of 6 and 10?

Answer: 30
Solution: $6 = 2^1 3^1$ and $10 = 2^1 5^1$. Both of these numbers share a factor of 2.
Thus, $LCM = 2^1 3^1 5^1 = 30$.

12. Find the mode of the set: {20, 20, 21, 19, 22, 19, 21, 21}.

Answer: 21
Solution: Rearranging the numbers: {19, 19, 20, 20, 21, 21, 21, 22}
21 appears the most number of times. Therefore, 21 is the mode.

13. How many yards are equivalent to 417 feet?

Answer: 139 (yards)
Solution: 3 feet is equivalent to 1 yard.
Therefore, 417 feet = (417 ÷ 3) yards = 139 yards

14. What is the units digit of the decimal representation of 3^7?

Answer: 7
Solution: $3^1 = 3$ ends in 3. $3^2 = 9$ ends in 9. $3^3 = 27$ ends in 7. $3^4 = 81$ ends in 1. $3^5 = 243$ ends in 3. The units digits follow a pattern for every 4 consecutive powers.
7 ÷ 4 = 1 R 3. Thus, 3^7 will have the same units digit as 3^3 which ends in 7.

15. What is the Greatest Common Factor (GCF) of 15 and 11?

Answer: 1
Solution: 11 is a prime number and 15 is not divisible by 11. Thus, the two numbers only share the factor 1.

16. The numerical value of a square's perimeter and area are equal. What is the measurement of its side length?

Answer: 4
Solution: $A = s^2$ and $P = 4s$.
Since $A = P$, $s^2 = 4s$. Dividing both sides by s: $s = 4$.

17. Find the value of $1 - 2 + 3 - 4 + 5 - 6 + 7 - 8 + 9 - 10$.

Answer: –5
Solution: Grouping pairs together: $(1 - 2) + (3 - 4) + (5 - 6) + (7 - 8) + (9 - 10) =$
$-1 - 1 - 1 - 1 - 1 = -5$.

18. What is the positive difference between 13 yards and 40 feet? Express your answer in inches.

Answer: 12 (inches)
Solution: 1 yard is equivalent to 3 feet. Thus, 13 yards = (13 × 3) feet = 39 feet.
The positive difference between 39 feet and 40 feet is (40 – 39) feet = 1 foot.
1 foot is equivalent to 12 inches.

19. April 5, 2014 was on a Saturday. What day of the week was April 25, 2014?

Answer: Friday
Solution: April 25 is 20 days after April 5. 20 days is one day less than 21 days which is equivalent to 3 weeks ($3 \times 7 = 21$). April 25 is 1 day less than 3 weeks after April 5 which is on a Saturday. Adding 3 weeks will keep the day of the week. Thus, 1 day before Saturday is Friday, which is the day of the week for April 25.

20. If $a + b = 6$ and $a - b = 4$, what is the value of a^2 ?

Answer: 25
Solution: Adding the two equations ($a + b = 6$) and ($a - b = 4$) gives $2a = 10$ or $a = 5$. Thus, $a^2 = 5^2 = 25$.

21. There are 9 total bicycles and tricycles in a parking lot. There are 21 wheels total. How many bicycles are there?

Answer: 6 (bicycles)
Solution: Assume that all 9 vehicles are bicycles. There are $9 \times 2 = 18$ wheels. This is $21 - 18 = 3$ wheels less than the total. This means that these 3 wheels are tricycles that need 1 more wheel. Thus, there are 3 tricycles and $9 - 3 = 6$ bicycles.

22. How many diagonals does a square have?

Answer: 2 (diagonals)

Solution: $$\mathrm{d} = \frac{n(n-3)}{2} = \frac{4(4-3)}{2} = \frac{4}{2} = 2$$

23. There are 2 blue marbles and 3 red marbles in a bag. If one marble is drawn at random, what is the probability that it is not blue? Express your answer as a common fraction.

Answer: $\frac{3}{5}$
Solution: There are $2 + 3 = 5$ total marbles. 3 of these marbles are not blue, which in this case is red. The probability is $\frac{3}{5}$.

24. How many factors does 45 have?

Answer: 6 (factors)

Solution: $45 = 3^2 5^1$. The number of factors is (2 + 1) × (1 + 1) = 3 × 2 = 6.

25. How many ways can the letters in the word "BAT" be arranged?

Answer: 6 (ways)
Solution: There are 3 different letters. Thus, there are 3! = 6 ways to arrange them.

26. What is the largest number less than 100 that is divisible by 5?

Answer: 95
Solution: 100 is divisible by 5 because it ends in 0. The largest number less than 100 is 100 – 5 = 95.

27. Shannon loves the number 5 so much that she finds the product of 5 × 5 × 5 × 5. What product does she calculate?

Answer: 625
Solution: (5 × 5) × (5 × 5) = 25 × 25 = 625

28. The probability that it rains on any given day is 10%. What is the probability that it does not rain on Tuesday? Express your answer as a common fraction.

Answer: $\frac{9}{10}$

Solution: The probability that it rains on any given day is 10% or $\frac{10}{100} = \frac{1}{10}$. Thus, the probability that it doesn't rain on any given day is $1 - \frac{1}{10} = \frac{9}{10}$.

29. Johnny writes down all the digits in the numbers 1 to 15, inclusive. How many digits does Johnny write down in total?

Answer: 21 (digits)
Solution: The one-digit numbers are from 1 to 9, inclusive, while the two-digit numbers are from 10 to 15, inclusive. There are (9 – 1) + 1 = 9 one-digit numbers and (15 – 10) + 1 = 6 two-digit numbers.
(9 × 1) + (6 × 2) = 9 + 12 = 21 digits.

30. How many cents are in 1 quarter, 2 dimes, 3 nickels and 4 pennies?

Answer: 64 (cents)
Solution: 1 quarter = 25 cents, 1 dime = 10 cents, 1 nickel = 5 cents, 1 penny = 1 cent.
$(1 \times 25) + (2 \times 10) + (3 \times 5) + (4 \times 1) = 25 + 20 + 15 + 4 = 64$ cents

31. Paul sees a movie that is x minutes long. If the movie starts at 3:25 PM and ends at 5:05 PM, what is the value of x?

Answer: 100
Solution: 2 hours after 3:25 PM is 5:25 PM. 20 minutes before 5:25 PM is 5:05 PM. Thus, the movie was 2 hours minus 20 minutes long. This is equivalent to $120 - 20 = 100$ minutes.

32. Julia's car costs $5.6 million while Rebecca's car costs $28.5 million. How much more does Rebecca's car cost than Julia's car in dollars?

Answer: $22,900,000
Solution: 28.5 million = 28,500,000 and 5.6 million = 5,600,000.
28,500,000 – 5,600,000 = 22,900,000.

33. Find the positive difference between 4102 and 2014.

Answer: 2088
Solution: $4102 - 2014 = 2088$

34. What is the greatest prime factor of 303?

Answer: 101
Solution: $303 = 3^1 101^1$. Thus, the greatest prime factor is 101.

35. What is the perimeter of a rectangle with area 19 if all of the side lengths are positive integers?

Answer: 40
Solution: $A = lw = 19$. Since the side lengths are positive integers and 19 is a prime number, $w = 1$ and $l = 19$. Thus, the perimeter is $P = 2l + 2w = 2 \times 19 + 2 \times 1 = 38 + 2 = 40$.

36. What is 10 increased by 200% ?

Answer: 30

Solution: 100% of 10 is 10. Thus, increasing it by 200% is equivalent to increasing it by 20.

10 + 20 = 30.

37. Find the range of the set: {3, 1, 4, 0, 9, 10, 5, 7, 8, 2, 6}

Answer: 10

Solution: Rearranging gives: {0, 1, 2, 3, 4, 5, 6, 7, 8, 9, 10}. The range is 10 – 0 = 10.

38. What is the side length of a cube with volume 125 units^3?

Answer: 5 (units)

Solution: $V = s^3 = 125$. $s = 5$ because $5^3 = 125$.

39. How many seconds are in 15 minutes?

Answer: 900 (seconds)

Solution: 1 minute = 60 seconds. Thus, 15 minutes = (15 × 60) seconds = 900 seconds.

40. What is the area of a triangle with sides 3, 4 and 5?

Answer: 6

Solution: The triangle is a right triangle because it is a pythagorean triple (3-4-5).

Thus, the area is $A = \frac{l_1 l_2}{2} = \frac{3 * 4}{2} = 6$

41. What is the measure of one of the angles in an equilateral triangle? Express your answer in degrees.

Answer: 60°

Solution: There are 180° in a triangle. Since each angle is equal in an equilateral triangle, the measure of one of the angles is 180 ÷ 3 = 60°.

42. How many days are in 192 hours?

Answer: 8 (days)

Solution: 1 day = 24 hours. Thus, 192 hours = (192 ÷ 24) days = 8 days

43. Find the circumference of a circle with area 4π. Express your answer in terms of π.

Answer: 4π
Solution: $A = \pi r^2 = 4\pi$. Thus, $r^2 = 4$ or $r = 2$.
$C = 2\pi r = 2\pi \times 2 = 4\pi$.

44. What is the remainder when 3 × 5 × 7 is divided by two?

Answer: 1
Solution: 3 × 5 × 7 is the product of 3 odd numbers. (Note: odd # × odd # × odd # = odd #) Thus, the product will be an odd number which will give a remainder of 1 when divided by two.

45. What is 20% of 60?

Answer: 12
Solution: $\frac{20}{100} * 60 = \frac{1}{5} * 60 = 12$

46. Express ⅜ as a decimal.

Answer: 0.375
Solution: ⅜ = 0.375 (See Formulas and Tips).
Another method is to convert the fraction so that the denominator is a power of 10.
$$\frac{3}{8} * \frac{125}{125} = \frac{375}{1000} = 0.375$$

47. What is the sum of the angles of a triangle? Express your answer in degrees.

Answer: 180°
Solution: n = 3.
$S = (n - 2)180° = (3 - 2)180° = 180°$

48. Solve. 6 × 5 × 4 × 3 × 2 × 1.

Answer: 720
Solution: 6 × 5 × 4 × 3 × 2 × 1 = 6! = 720.

49. What is the hypotenuse of a right triangle with legs 5 and 12?

Answer: 13
Solution: The pythagorean triple is 5-12-13. Thus, the hypotenuse is 13. Using the pythagorean theorem: $\sqrt{5^2 + 12^2} = \sqrt{25 + 144} = \sqrt{169} = 13$

50. How many ounces are in two pounds?

Answer: 32 (ounces)
Solution: 1 pound = 16 ounces. Thus, 2 pounds = (2 × 16) ounces = 32 ounces.

51. A toy car costs 95 cents. What is the most number of toy cars that Todd can buy with his $5.00 allowance?

Answer: 5 (toy cars)
Solution: $5.00 = 500 cents.
500 ÷ 95 = 5 R 25. Thus, he can only buy a maximum of 5 toy cars.

52. Sally can build eight cars in one hour. How many cars can she build in 30 minutes?

Answer: 4 (cars)
Solution: 30 minutes = ½ hour. Thus, she can make ½ as many cars as she would make in one hour. ½ × 8 = 4 cars.

53. Solve. 2 + 4 + 6 + 8 + 10 + 12 + 14 + 16 + 18 + 20.

Answer: 110
Solution: $n = 10$, $a_1 = 2$ and $a_{10} = 20$

Sum of Arithmetic Sequence: $\frac{10(2+20)}{2} = \frac{10 * 22}{2} = 10 * 11 = 110$

54. Find the product of all of the digits in the number 125,783,401,839,824.

Answer: 0
Solution: The middle digit is 0. Thus, the product of all the digits will be 0.

55. A fair die is rolled. What is the probability that the number rolled is a multiple of two? Express your answer as a common fraction.

Answer: ½
Solution: There are 6 possibilities (1, 2, 3, 4, 5, 6) and 3 of them are multiples of two (2, 4, 6). Thus, the probability is $\frac{3}{6} = \frac{1}{2}$.

56. A bus in Mathville leaves the station every 11 minutes. If a bus just left the station at 10:54 PM, what time will the next bus leave the station?

Answer: 11:05 PM
Solution: 11 minutes after 10:54 PM is 11:05 PM.

57. What is the Least Common Multiple (LCM) of 1, 2, and 3?

Answer: 6
Solution: 1, 2 and 3 do not share any factors beside 1. Therefore, LCM = 1 × 2 × 3 = 6

58. A customer purchases a newspaper for 99 cents with $1.50. What amount of change should the customer receive? Express your answer in cents.

Answer: 51 (cents)
Solution: $1.50 = 150 cents. Thus, the customer should receive 150 – 99 = 51 cents.

59. How many seconds are in 2 hours?

Answer: 7200 (seconds)
Solution: 1 hour = 60 minutes. 1 minute = 60 seconds.
Thus, 2 hours = (2 × 60) minutes = 120 minutes = (120 × 60) seconds = 7200 seconds.

60. What is the smallest prime factor of 24?

Answer: 2
Solution: The smallest prime number is 2. Since 24 is even, it is divisible by 2. Thus, its smallest prime factor is 2.

61. Two coins are flipped. What is the probability that zero heads show up? Express your answer as a common fraction.

Answer: ¼

Solution: There are $2^2 = 4$ possible outcomes. The probability that zero heads show up is when the outcome is TT (Tails-Tails). Thus, the probability is ¼.

62. 10 yards is equivalent to x inches. What is the value of x?

Answer: 360

Solution: 1 yard = 3 feet. 1 foot = 12 inches.
Thus, 10 yards = (10 × 3) feet = 30 feet = (30 × 12) inches = 360 inches.

63. Find the largest counting number less than 8 × 6 × 4 × 2.

Answer: 383

Solution: (8 × 6) × (4 × 2) = 48 × 8 = 384. The largest counting number less than 384 is 384 – 1 = 383.

64. Sarah makes $20 an hour working at a convenience store. If she works 80 hours a month, how much money does she make each month? Express your answer in dollars.

Answer: $1600

Solution: 80 × $20 = $1600

65. How many diagonals does a convex quadrilateral have?

Answer: 2 (diagonals)

Solution: A quadrilateral has 4 sides. $\mathrm{d} = \frac{n(n-3)}{2} = \frac{4(4-3)}{2} = \frac{4}{2} = 2$

66. Find the units digit of 6^6.

Answer: 6

Solution: $6^1 = 6$ ends in 6. $6^2 = 36$ ends in 6. Thus, 6^n ends in 6 for all positive integers n.

67. Find the value of $20^2 - 14^2$.

Answer: 204

Solution: Using the difference of two squares:
$20^2 - 14^2 = (20 - 14)(20 + 14) = 6 \times 34 = 204$.

68. James took a lunch break from work. He left work at 11:00 AM and returned at 12:37 PM. How long was James' lunch break in minutes?

Answer: 97 (minutes)
Solution: 11:00 AM to 12:37 PM is 1 hours and 37 minutes. This is equivalent to 60 + 37 = 97 minutes.

69. What is the diagonal of a rectangle with side lengths 9 and 12?

Answer: 15
Solution: The diagonal of a rectangle forms a right triangle with the length and the width of the rectangle. The diagonal is the hypotenuse of the right triangle with legs 9 and 12. The pythagorean triple 9-12-15 (which is 3-4-5 multiplied by a factor of 3) where the hypotenuse equals 15. This can also be found using the pythagorean theorem:
$\sqrt{9^2 + 12^2} = \sqrt{81 + 144} = \sqrt{225} = 15$

70. How many prime numbers are between 30 and 40?

Answer: 2 (numbers)
Solution: 31 and 37 are the two prime numbers between 30 and 40.

71. How many factors does the number 25 have?

Answer: 3 (factors)
Solution: $25 = 5^2$. Thus, it has (2 + 1) = 3 factors.

72. A die is rolled. What is the probability that the number rolled is a 5? Express your answer as a common fraction.

Answer: $\frac{1}{6}$
Solution: There are 6 possibilities (1, 2, 3, 4, 5, 6) and 1 of them is a 5. Thus, the probability is $\frac{1}{6}$.

73. How many feet are in 5 miles?

Answer: 26,400 (feet)
Solution: 1 mile = 5280 feet. Thus, 5 miles = (5 × 5280) feet = 26,400 feet.

74. Express ⅗ as a percent.

Answer: 60%
Solution: ⅗ = 60%. Another way is to make the denominator equal to 100:
$\frac{3}{5} * \frac{20}{20} = \frac{60}{100}$ or 60%.

75. What is the value of a if $(a - 2014) = 2014$?

Answer: 4028
Solution: Add 2014 to both sides of the equation: $a = 2014 + 2014 = 4028$.

76. Find the surface area of a cube with side length 4.

Answer: 96
Solution: s = 4. Surface Area = $6s^2 = 6 \times 4^2 = 6 \times 16 = 96$.

77. A deluxe toy costs $15.49, while a premium toy costs $29.79. How much more does the premium toy cost than the deluxe toy? Express your answer in dollars.

Answer: $14.30
Solution: $29.79 – $15.49 = $14.30.

78. How many pints are in 3 quarts?

Answer: 6 (pints)
Solution: 1 quart = 2 pints. Thus, 3 quarts = (3 × 2) pints = 6 pints.

79. Find the median of the set: {9, 12, 13, 5, 1}.

Answer: 9
Solution: Rearranging the set: {1, 5, 9, 12, 13}. Thus, the median is 9.

80. Solve. 2014 × 2 – 1007 × 4.

Answer: 0
Solution: (2014 × 2) – (1007 × 4) = 4028 – 4028 = 0.

[This page is intentionally left blank]

2015
Answer Keys and Solutions

2015 Individual Round Solutions

Answer Key

1. c) 51
2. c) 27 ft^3
3. d) 29
4. b) ⅔
5. d) 13
6. e) 49 in^2
7. d) 24
8. d) 111
9. c) 14
10. e) 19
11. b) 30
12. a) $7
13. a) 5
14. b) 48 $meters^2$
15. c) ½
16. b) 11
17. a) 4320 minutes
18. c) 120π
19. d) 8
20. c) 10
21. 43
22. 24 (feet)
23. 148
24. 8 (pieces)
25. 300 (miles)
26. 70
27. 54 ($units^2$)
28. 9 (diagonals)
29. $1.50
30. 29

1. Find the 17th positive multiple of 3.

Answer: c) 51

Solution: The 17th positive multiple of 3 is: $3 \times 17 = 51$.

2. What is the volume of a cube with side length 3 feet?

Answer: c) 27 ft^3

Solution: $V = s^3 = 3^3 = 27 \text{ ft}^3$

3. Find the next number in the arithmetic sequence: {1, 5, 9, 13, 17, 21, 25, __}

Answer: d) 29

Solution: The common difference is $5 - 1 = 4$. Thus, the next number is $25 + 4 = 29$.

4. A bag contains 1 red, 1 green and 1 blue marble. Jane randomly selects one marble from the bag. What is the probability that Jane draws a non-red marble?

Answer: b) ⅔

Solution: There are $(1 + 1 + 1) = 3$ total marbles. There are 2 non-red marbles (1 green and 1 blue). Thus, the probability that she draws a non-red marble is ⅔.

5. If Jack wrote all of the numbers from 1 to 30, how many times would he write the digit 2?

Answer: d) 13

Solution: The digit 2 appears in the units digits and the tens digit. It appears in the units digit 3 times (2, 12 and 22) and appears in the tens digit 10 times (20 to 29).
Thus, he writes the digit 2: $(3 + 10) = 13$ times.

6. A square has perimeter 28 inches. What is the area of the square?

Answer: e) 49 in^2

Solution: P = 4s = 28. s = 28 ÷ 4 = 7 in.
$A = s^2 = 7^2 = 49\ in^2$.

7. Find the sum of the factors of 14.

Answer: d) 24

Solution: $14 = 2^1 7^1$
Sum of Factors = $(2^0 + 2^1) \times (7^0 + 7^1) = (1 + 2) \times (1 + 7) = 3 \times 8 = 24$

8. Given the set: {1, 11, 111, 1111, 11111}. Find the median.

Answer: d) 111

Solution: There are 5 numbers. Thus, the median is the middle number which is 111.

9. How many numbers between 1 and 71 are multiples of 5?

Answer: c) 14

Solution: 71 ÷ 5 = 14 R 1. Thus, there are 14 multiples of 5 between 1 and 71.

10. How many weeks are in 133 days?

Answer: e) 19

Solution: There are 7 days in a week.
Thus, 133 days = (133 ÷ 7) weeks = 19 weeks

11. The ratio of boys to girls in a classroom is 1:2. If there are 15 boys in the classroom, how many girls are there?

Answer: b) 30

Solution: The ratio implies that there are 1 boy for every 2 girls.
Thus, there are $15 \times 2 = 30$ girls.

12. One newspaper and two magazines cost \$11. Two newspapers and one magazine cost \$10. How much does one newspaper and one magazine cost?

Answer: a) \$7

Solution: Adding the two equations yield: 3 newspapers and 3 magazines = \$11 + \$10 = \$21.
Dividing both sides by 3: 1 newspaper and 1 magazine = \$21 ÷ 3 = \$7.

13. Find the value of a if $a + b = 16$, $ab = 55$, and $a < b$, where a and b are counting numbers.

Answer: a) 5

Solution: $ab = 55 = 5^1 11^1$. Since a and b are counting numbers, there are limited number of possibilities. $a = 1, b = 55$ and $a = 5, b = 11$ are the only possibilities because 55 has 4 factors and $a < b$.
The only solution that $a + b = 16$ is when $a = 5$ and $b = 11$. Thus, $a = 5$.

14. What is the area of a rectangle with length 8 meters and diagonal 10 meters?

Answer: b) 48 meters^2

Solution: The length, the width and the diagonal of a rectangle form a right triangle. Thus, the width can be found using the pythagorean theorem and the difference of two squares.

$$w = \sqrt{10^2 - 8^2} = \sqrt{(10-8)(10+8)} = \sqrt{2 * 18} = \sqrt{36} = 6$$

Now, the area of the rectangle can be found: $A = lw = 8 \times 6 = 48$.
(Note: The pythagorean triple 6-8-10 which is 3-4-5 multiplied by 2 can also be used to find the width.)

15. Three fair coins are flipped. What is the probability that more heads come up than tails?

Answer: c) ½

Solution: There is an odd number of coins. Thus, there will either be more heads or more tails each time the three coins are flipped. Since they are fair coins, the probability that there will be more heads is the same as the probability that there will be more tails. Since these two probabilities sum to 1, the probability that there are more heads or the probability that there are more tails is ½.

16. Using only pennies, nickels, dimes and quarters, what is the least number of coins necessary to pay a $1.44 toll?

Answer: b) 11

Solution: Since the least number of coins is to be found, the coins with the largest value should be maximized. Thus, the quarter (worth 25 cents) should be maximized. Afterwards, the dime, then the nickel, and finally the penny.
$1.44 = 144 cents
144 ÷ 25 = 5 R 19 (5 quarters)
19 ÷ 10 = 1 R 9 (1 dime)
9 ÷ 5 = 1 R 4 (1 nickel and 4 pennies)
Thus, the least number of coins is (5 + 1 + 1 + 4) = 11 coins.

17. How many minutes are in 3 days?

Answer: a) 4320 minutes

Solution: 1 day = 24 hours. 1 hour = 60 minutes.
3 days = (3 × 24) hours = 72 hours = (72 × 60) minutes = 4320 minutes

18. A circle has area 144π. What is the positive difference of the numerical value between the area and the circumference of the circle?

Answer: c) 120π

Solution: $A = \pi r^2 = 144\pi$. $r^2 = 144$. $r = 12$. (Note: $12^2 = 144$)
$C = 2\pi r = 2\pi \times 12 = 24\pi$.
Positive Difference = $A - C = 144\pi - 24\pi = 120\pi$.

19. How many prime numbers are between 23 and 53, inclusive?

Answer: d) 8

Solution: The prime numbers between 23 and 53, inclusive, are 23, 29, 31, 37, 41, 43, 47 and 53. Thus, there are 8 prime numbers between 23 and 53, inclusive.

20. Find the number of factors for 48.

Answer: c) 10

Solution: $48 = 2^4 3^1$. Thus, the number of factors is $(4 + 1) \times (1 + 1) = 5 \times 2 = 10$ factors.

21. Given a set: {1, 13, 25, 37, 49, 61, 73, 85}. Find the mean.

Answer: 43

Solution: The 8 numbers in the set are already in order. In addition, the numbers follow an arithmetic sequence where the common difference is $13 - 1 = 12$.
Thus, the mean of the set will be equivalent to the median of the set.
The median is the average of 37 and 49: $(37 + 49) \div 2 = 86 \div 2 = 43$. The mean is 43.

22. A rectangle has length and width that are whole numbers. If the area of the rectangle is 11 square feet, what is its perimeter? Express your answer in feet.

Answer: 24 (feet)

Solution: $A = lw = 11$. Since 11 is a prime number and the length and width are whole numbers, $l = 11$ and $w = 1$.
Thus, the perimeter is $P = 2l + 2w = (2 \times 11) + (2 \times 1) = 22 + 2 = 24$ feet.

23. What is the surface area of a rectangular prism with length 6, width 5, and height 4?

Answer: 148

Solution: The surface area of a rectangular prism is:
Surface Area = $2(lw + wh + lh) = 2(6 \times 5 + 5 \times 4 + 6 \times 4) = 2(30 + 20 + 24) = 2 \times 74 = 148$

24. Patty was giving away a cake with 24 equal pieces. Amanda ate half of the cake. Soon after, Billy ate one third of the remainder of the cake. Finally, Charlie ate the rest of the cake. How many pieces did Charlie eat?

Answer: 8 (pieces)

Solution: If Amanda ate half of the cake, there is $24 \div 2 = 12$ pieces left.
Thus, Billy ate $12 \div 3 = 4$ pieces of the cake.
Therefore, Charlie ate $(12 - 4) = 8$ pieces.

25. Joel drives at a speed of 100 miles per hour. How many miles will he drive in 180 minutes?

Answer: 300 (miles)

Solution: 60 minutes = 1 hour. Thus, 180 minutes is equivalent to 3 hours ($3 \times 60 = 180$).
Using Distance = Rate × Time,
Distance = (100 miles per hour) × (3 hours) = 300 miles.

26. What is the Least Common Multiple (LCM) of 14 and 35?

Answer: 70

Solution: $14 = 2^1 7^1$ and $35 = 5^1 7^1$. 14 and 35 share the factor 7.
Thus, the LCM is $2^1 5^1 7^1 = 70$.

27. A cube has volume 27 units^3. What is the surface area of the cube?

Answer: 54 (units^2)

Solution: $V = s^3 = 27$. $s = 3$. (Note: $3^3 = 27$)
Surface Area = $6s^2 = 6 \times 3^2 = 6 \times 9 = 54$ units^2.

28. How many diagonals does a hexagon have?

Answer: 9 (diagonals)

Solution: A hexagon has 6 sides. $d = \frac{n(n-3)}{2} = \frac{6(6-3)}{2} = \frac{6*3}{2} = \frac{18}{2} = 9$.

29. Rocky and Halle are paying their bill at Maui Math Restaurant. They also have to pay a tip that is 15% of the original bill. If the total amount including tip came to $11.50, how much was the tip?

Answer: $1.50

Solution: If the total amount with the tip totaled to $11.50, this must be (100 + 15)% = 115% or 1.15 of the original price.
Original Bill × Percentage (with tip) = Total Bill

$$\text{Original} = \frac{\text{Total Bill}}{\text{Percentage (with tip)}} = \frac{11.50}{1.15} = 10$$

This means that the original bill was $10.00 and the tip was ($11.50 – $10.00) = $1.50.

30. What is the 10th smallest prime number?

Answer: 29

Solution: The first 10 prime numbers are 2, 3, 5, 7, 11, 13, 17, 19, 23, 29. Thus, 29 is the 10th smallest prime number.

2015 Team Round Solutions

Answer Key

1. 42
2. 87
3. 21 (marbles)
4. 7 (ways)
5. 10%
6. 10 (ways)
7. 25π (units^2)
8. 100 (students)
9. 251
10. 36 (handshakes)

1. If a + b + c = 21, a + b = 14 and a + c = 15, what is the product of b and c?

Answer: 42

Solution: Let (a + b + c = 21) be equation (1).
Let (a + b = 14) be equation (2).
Let (a + c = 15) be equation (3).
Subtracting (2) from (1) gives: (a + b + c) – (a + b) = c = 21 – 14 = 7
Subtracting (3) from (1) gives: (a + b + c) – (a + c) = b = 21 – 15 = 6
Thus, b × c = 6 × 7 = 42.

2. A list of three different positive integers is written in numerical order from least to greatest. The sum of the numbers is 90 and the median is 2. What is the greatest number in the list?

Answer: 87

Solution: If the median is 2, the smallest number must be 1 because the three positive integers are different. The first two numbers are 1 and 2.
The greatest number is therefore, (90 – 1 – 2) = 87.

3. A bag has 2 red marbles, 5 blue marbles, 7 green marbles and 11 black marbles. What is the least number of marbles that Troy has to take out to guarantee that he chose at least 1 blue marble?

Answer: 21 (marbles)

Solution: In order to guarantee that he chose at least 1 blue marble, he must take out all the non-blue marbles and one blue marble.
Red + Green + Black + 1 (Blue) = 2 + 7 + 11 + 1 = 21 marbles.

4. Using only quarters, dimes, and nickels, how many different ways are there to make 40 cents? (Hint: You do not need to use all of the different coins.)

Answer: 7 (ways)

Solution:
Using casework, there are 7 ways.

Nickels (5 cents)	Dimes (10 cents)	Quarters (25 cents)	Total = 40 cents
1	1	1	40 cents
3	0	1	40 cents
0	4	0	40 cents
2	3	0	40 cents
4	2	0	40 cents
6	1	0	40 cents
8	0	0	40 cents

5. A number is randomly chosen from 1 to 100, inclusive. What is the probability that the number is divisible by 2 and 5? Express your answer as a percent.

Answer: 10%

Solution: A number that is divisible by 2 and 5 is divisible by 2 × 5 = 10 since 2 and 5 are both prime numbers.
There are (100 ÷ 10) = 10 numbers that are divisible by 10 from 1 to 100, inclusive.
Thus, the probability is $\frac{10}{100} * 100 = 10 \text{ percent}$ or 10%.

6. How many ways can the letters in the word MAMMA be arranged?

Answer: 10 (ways)

Solution: There are 5 letters, hence, there are 5! ways to arrange them. However, there are 3 M's and 2 A's which can be arranged 3! and 2! ways, respectively.

Thus, there are $\frac{5!}{2! * 3!} = \frac{120}{6 * 2} = \frac{120}{12} = 10$ ways.

7. A circle is inscribed in a square with side length 10 units. Find the area of the circle in terms of π.

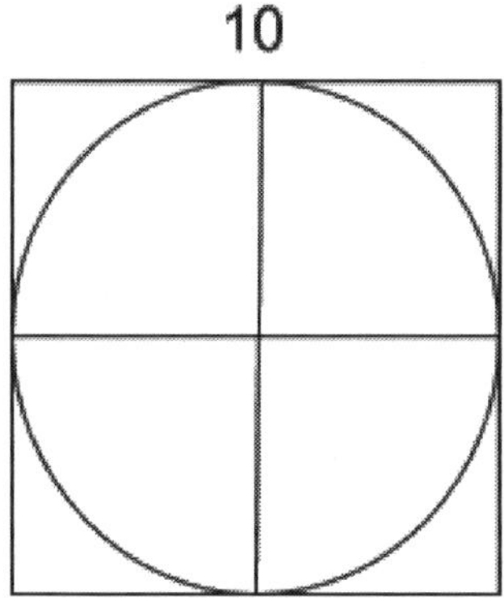

Answer: 25π

Solution: Divide up the square and the circle using two lines, the diameter of the circle is the same length as the side length of the square, which is 10. Thus, d = 2r = 10 and r = 5. Therefore, $A = \pi r^2 = \pi \times 5^2 = 25\pi$.

8. The ratio of boys to girls in a classroom is 9 to 1. However, after 20 boys leave the room, the new ratio of boys to girls is 7 to 1. How many total students were there in the beginning?

Answer: 100 (students)

Solution: Originally, there were 9 boys to 1 girl. After 20 boys left, there were 7 boys to 1 girl. Since the number of girls is constant, the number of girls can be found.
For each girl, 2 boys left (9 – 7 = 2). Thus, there are (20 ÷ 2) = 10 girls.
Using the original ratio, this means that there were originally (9 × 10) = 90 boys.
Thus, there were (90 + 10) = 100 students in the beginning.

9. Find the sum of the prime numbers between 20 and 50.

Answer: 251

Solution: The prime numbers between 20 and 50 are 23, 29, 31, 37, 41, 43, and 47. Thus, the sum is (23 + 29) + (31 + 37) + (41 + 43 + 47) = 52 + 68 + 131 = 251.

10. Andy, Bill, Carrie, David, Eddy, Freddie, Greg, Hannah and Isabella are at a party. At the beginning of the party, they shake hands with each other. If each person shook hands with another person only once, how many handshakes were there in total?

Answer: 36 (handshakes)

Solution: There are 9 people at this party. Each person shakes hands with (9 – 1) = 8 people.
Therefore, 9 × 8 = 72. This number is divided by 2 because Andy shaking hands with Bill is the same as Bill shaking hands with Andy.

72 ÷ 2 = 36 handshakes.

2015 Countdown Round Solutions

1. 4 (numbers)
2. 21
3. 3
4. 100 (pennies)
5. $\frac{1}{8}$
6. 8π
7. 73 (integers)
8. 42
9. 4567
10. 512
11. 24 (cups)
12. 6
13. 40 (people)
14. 120 (miles)
15. 10,800
16. $1.51
17. 11 (multiples)
18. 41,976
19. $80.00
20. 4
21. 3
22. 5
23. 120
24. 75
25. 151 (minutes)
26. 28
27. 98
28. 90°
29. 64π
30. 0
31. 2 (diagonals)
32. 0 (multiples)
33. 6
34. $2.00
35. 11 (digits)
36. 1001
37. 10
38. 26
39. 0
40. 30
41. 30
42. 3520 (yards)
43. 56 (ounces)
44. $\frac{1}{6}$
45. 6 (ways)
46. 5
47. 97
48. 40%
49. 2020
50. 11
51. 36 (feet^2)
52. 10,080
53. $2.06
54. 10 (inches)
55. 30
56. 38 (integers)
57. 125 (nickels)
58. 0
59. 40 (pints)
60. 195
61. 576
62. 7:26 AM
63. 0
64. 31
65. 100
66. 0
67. 96
68. 72
69. 1
70. 153
71. 24 (ways)
72. 0
73. 75
74. 540°
75. 7 (miles)
76. 7:30 PM
77. 80
78. 0
79. 11
80. 4031 (integers)

1. How many prime numbers are between 1 and 10?

Answer: 4 (numbers)
Solution: The 4 prime numbers between 1 and 10 are 2, 3, 5 and 7.

2. If $a + b = 10$ and a is 7, what is the product of a and b?

Answer: 21
Solution: If $a = 7$, then $7 + b = 10$. Subtracting 7 from both sides gives $b = 10 - 7 = 3$. Thus, the product of a and b is $a \times b = 7 \times 3 = 21$.

3. Find the mode of the set: {1, 3, 3, 9, 10}.

Answer: 3
Solution: 3 appears the most out of all the numbers in the set. Thus, 3 is the mode.

4. How many pennies are needed to exchange for 20 nickels?

Answer: 100 (pennies)
Solution: 1 nickel = 5 pennies. Thus, 20 nickels = (20 × 5) pennies = 100 pennies

5. Three fair coins are flipped. What is the probability of getting 0 tails? Express your answer as a common fraction.

Answer: $\frac{1}{8}$
Solution: There are $2^3 = 8$ possible outcomes. Only one of the outcomes has 0 tails (Head-Head-Head). Thus, the probability of getting 0 tails is $\frac{1}{8}$.

6. What is the circumference of a circle with area 16π? Express your answer in terms of π.

Answer: 8π
Solution: $A = \pi r^2 = 16\pi$. $r^2 = 16$. $r = 4$.
$C = 2\pi r = 2\pi \times 4 = 8\pi$.

7. How many integers are between 15 and 87, inclusive?

Answer: 73 (integers)
Solution: There are $(87 - 15) + 1 = 73$ integers between 15 and 87, inclusive.

8. Find the value of 3 + 4 + 5 + 6 + 7 + 8 + 9.

Answer: 42

Solution: $S_n = \frac{n(a_1 + a_n)}{2} = \frac{7(3+9)}{2} = \frac{7 * 12}{2} = 42$

9. What is 50% of 9134?

Answer: 4567

Solution: 50% = ½. Thus, 50% of 9134 = 9134 ÷ 2 = 4567

10. What is the volume of a cube with side length 8?

Answer: 512

Solution: $V = s^3 = 8^3 = 512$.

11. How many cups are in 1.5 gallons?

Answer: 24 (cups)

Solution: 1 gallon = 4 quarts. 1 quart = 2 pints. 1 pint = 2 cups.
Thus, 1.5 gallons = (1.5 × 4) quarts = 6 quarts = (6 × 2) pints = 12 pints = (12 × 2) cups = 24 cups.

12. What is the units digit of 6^4?

Answer: 6

Solution: 6^n always ends in 6 for any positive integer *n*. Thus, 6^4 ends in 6.

13. The number of people in a room was 50. After one hour, 20% of them left the room. What is the remaining number of people in the room?

Answer: 40 (people)

Solution: 20% of 50 left. This is equivalent to $\frac{20}{100} * 50 = 10$.
If 10 people left, there are (50 – 10) = 40 people remaining.

14. Jack rides his bike at a constant rate of 5 miles per hour. What distance can he travel in one day assuming that he does not stop biking? Express your answer in miles.

Answer: 120 (miles)
Solution: There are 24 hours in a day. Thus, using Distance = Rate × Time:
Distance = (5 miles per hour) × (24 hours) = 120 miles

15. How many seconds are in 3 hours?

Answer: 10,800 (seconds)
Solution: 1 hours = 60 minutes. 1 minute = 60 seconds.
3 hours = (3 × 60) minutes = 180 minutes = (180 × 60) seconds = 10,800 seconds

16. Larry bought a toy car at a store for $8.49. What change did he receive if he paid with $10.00? Express your answer in dollars.

Answer: $1.51
Solution: $10.00 – $8.49 = $1.51

17. How many multiples of 10 are between 21 and 133?

Answer: 11 (multiples)
Solution: The first multiple of 10 between 21 and 133 is 30 and the last multiple of 10 is 130.
30 = 3 × 10 and 130 = 13 × 10. Thus, there are (13 – 3) + 1 = 11 multiples.

18. Find the positive difference between 12345 and 54321.

Answer: 41,976
Solution: 54,321 – 12,345 = 41,976

19. If James makes $10 per hour and Jack makes twice as much as James, how much money does Jack make in 4 hours?

Answer: $80.00
Solution: If Jack makes twice as much, he makes $10 × 2 = $20 per hour.
Thus, if he works for 4 hours, he makes $20 per hour × 4 hours = $80.

20. Find the mean of the set: {1, 2, 3, 4, 5, 6, 7}.

Answer: 4
Solution: The set is an arithmetic sequence. The mean is equivalent to the median. The median is 4. Therefore, the mean is 4.

21. What is the greatest prime factor of 24?

Answer: 3
Solution: $24 = 2^3 3^1$. Thus, 3 is its largest prime factor.

22. Find the units digit of 5^8.

Answer: 5
Solution: 5^n always ends in 5 for any positive integer *n*. Thus, 5^8 ends in 5.

23. What is the value of 5 × 4 × 3 × 2 × 1?

Answer: 120
Solution: 5! = 5 × 4 × 3 × 2 × 1 = 120.

24. Find the sum of the integers from 10 to 15, inclusive.

Answer: 75
Solution: There are (15 – 10) + 1 = 6 numbers.

$$S_n = \frac{n(a_1 + a_n)}{2} = \frac{6(10+15)}{2} = \frac{6 * 25}{2} = 75$$

25. A movie starts at 10:00 PM and ends at 12:31 AM. How long is the movie in minutes?

Answer: 151 (minutes)
Solution: The movie last 2 hours and 31 minutes. 1 hour = 60 minutes.
Thus, 2 hours and 31 minutes = (2 × 60) + 31 = 120 + 31 = 151 minutes

26. Find the sum of all the factors of 12.

Answer: 28
Solution: $12 = 2^2 3^1$. Thus, the sum of the factors is $(2^0 + 2^1 + 2^2) \times (3^0 + 3^1) = 7 \times 4 = 28$.

27. What is 10% of 10% of 9800?

Answer: 98

Solution: $\frac{10}{100} * \frac{10}{100} * 9800 = \frac{1}{10} * \frac{1}{10} * 9800 = 98$

28. What is one of the angles in a rectangle? Express your answer in degrees.

Answer: 90°

Solution: Each angle in a rectangle is a right angle which is 90 degrees.

29. What is the area of a circle with radius 8? Express your answer in terms of π.

Answer: 64π

Solution: $A = \pi r^2 = \pi \times 8^2 = 64\pi$.

30. Find the value of 1 + 1 – 1 – 1 + 1 + 1 – 1 – 1.

Answer: 0

Solution: (1 + 1 – 1 – 1) + (1 + 1 – 1 – 1) = (2 – 2) + (2 – 2) = 0 + 0 = 0

31. How many diagonals does a rhombus have?

Answer: 2 (diagonals)

Solution: A rhombus has 4 sides. Thus, $d = \frac{n(n-3)}{2} = \frac{4(4-3)}{2} = \frac{4 * 1}{2} = 2$

32. How many multiples of 23 are between 48 and 55?

Answer: 0 (multiples)

Solution: The first few multiples of 23 are 23, 46, and 69. 46 is less than 48 and 69 is greater than 55. Thus, there are no multiples of 23 that are between 48 and 55.

33. Find the value of 5 × 4 × 3 × 2 × 1 × 0 + 3 × 2.

Answer: 6

Solution: (5 × 4 × 3 × 2 × 1 × 0) + (3 × 2) = 0 + 6 = 6

34. How many dollars are in 5 quarters, 5 nickels and 5 dimes?

Answer: \$2.00
Solution: 1 quarter = 25 cents. 1 nickel = 5 cents. 1 dime = 10 cents.
5 quarters + 5 nickels + 5 dimes = (5 × 25) + (5 × 5) + (5 × 10) = 125 + 25 + 50 = 200 cents
200 cents = \$2.00.

35. How many digits does it take to write out all the integers from 1 to 10, inclusive?

Answer: 11 (digits)
Solution: There are 9 integers that are single-digits (1 to 9). 10 is the only integer with 2 digits. Thus, there is a total of (9 + 2) = 11 digits.

36. Find the product: 7 × 11 × 13.

Answer: 1001
Solution: 7 × 11 × 13 = 1001 is a useful fact to know.

37. If a × b = 24 and b = 4, what is the value of a + b?

Answer: 10
Solution: If b = 4, a = 24 ÷ 4 = 6. Thus, a + b = 6 + 4 = 10.

38. What is the hypotenuse of a right triangle with legs 10 and 24?

Answer: 26
Solution: The pythagorean triple is 5-12-13 multiplied by a factor of 2 to give 10-24-26. Thus, the hypotenuse is 26. The pythagorean theorem can also be used to find the hypotenuse: $\sqrt{10^2 + 24^2} = \sqrt{100 + 576} = \sqrt{676} = 26$.

39. Find the product of all the integers between −15 and 15.

Answer: 0
Solution: 0 is an integer between −15 and 15. Thus, the product of all the integers between −15 and 15 is 0.

40. If a = 6, what is the value of $a^2 - a$?

Answer: 30
Solution: $a^2 - a = 6^2 - 6 = 36 - 6 = 30$.

41. What is the Least Common Multiple (LCM) of 10 and 15?

Answer: 30
Solution: $10 = 2^1 5^1$ and $15 = 3^1 5^1$. 10 and 15 share the factor 5.
Thus, the LCM is $2^1 3^1 5^1 = 30$.

42. How many yards are in 2 miles?

Answer: 3520 (yards)
Solution: 1 mile = 5280 feet. 1 yard = 3 feet.
Thus, 2 miles = (2 × 5280) feet = 10560 feet = (10560 ÷ 3) yards = 3520 yards

43. How many ounces are in 3.5 pounds?

Answer: 56 (ounces)
Solution: 1 pound = 16 ounces.
Thus, 3.5 pounds = (3.5 × 16) ounces = 56 ounces

44. A regular die is rolled. What is the probability that the number rolled is a multiple of 4? Express your answer as a common fraction.

Answer: $\frac{1}{6}$
Solution: There are 6 possibilities (1, 2, 3, 4, 5, 6) and only 1 of them is a multiple of 4 (4). Thus, the probability is $\frac{1}{6}$.

45. How many ways can the letters in the word "CAR" be arranged?

Answer: 6 (ways)
Solution: There are 3 different letters. Thus, it can be arranged 3! = 6 ways.

46. Find the value of 10 – 9 + 8 – 7 + 6 – 5 + 4 – 3 + 2 – 1.

Answer: 5
Solution: (10 – 9) + (8 – 7) + (6 – 5) + (4 – 3) + (2 – 1) = 1 + 1 + 1 + 1 + 1 = 5

47. What is the largest prime number less than 100?

Answer: 97
Solution: The largest prime number less than 100 can be found by testing the largest odd numbers less than 100. 99 is divisible by 3, hence it is not prime. 97 is not divisible by any numbers except 1 and 97. Thus, 97 is the largest prime number less than 100.

48. Express ⅖ as a percent.

Answer: 40%
Solution: ⅖ = 40%. It can also be found by making the denominator equal to 100:
$\frac{2}{5} * \frac{20}{20} = \frac{40}{100}$ which is equivalent to 40%.

49. What is the value of 2015 + 20 – 15?

Answer: 2020
Solution: 2015 + (20 – 15) = 2015 + 5 = 2020

50. Find the positive difference between the smallest prime number and the largest prime number less than 15.

Answer: 11
Solution: The smallest prime number is 2. The largest prime number less than 15 is 13. Thus, the positive difference is 13 – 2 = 11.

51. What is the area of a square with perimeter 24 feet? Express your answer in square feet.

Answer: 36 (feet^2)
Solution: P = 4s = 24. s = 24 ÷ 4 = 6.
A = $s^2 = 6^2 = 36$ feet^2

52. How many minutes are in 7 days?

Answer: 10,080 (minutes)
Solution: 1 day = 24 hours. 1 hour = 60 minutes.
Thus, 7 days = (7 × 24) hours = 168 hours = (168 × 60) minutes = 10,080 minutes

53. If Jackie buys 3 dolls, each cost 98 cents, with a \$5.00 bill, how much should she receive in change? Express your answer in dollars.

Answer: \$2.06
Solution: Let a doll cost 1 dollar minus 2 cents, then the total for 3 dolls is \$3.00 minus 6 cents or \$2.94.
The change is: \$5.00 – \$2.94 = \$2.06.

54. Find the diagonal of a rectangle with length 8 inches and width 6 inches. Express your answer in inches.

Answer: 10 (inches)
Solution: The length, the width and the diagonal of a rectangle form a right triangle. Thus, the pythagorean theorem can be applied to find the diagonal.
$\sqrt{6^2+8^2} = \sqrt{36+64} = \sqrt{100} = 10$. The pythagorean triple that matches is 3-4-5 multiplied by a factor of 2 to give 6-8-10 where 10 is the hypotenuse or the diagonal.

55. What is 20% of 50% of 300?

Answer: 30
Solution: $\frac{20}{100} * \frac{50}{100} * 300 = \frac{1}{5} * \frac{1}{2} * 300 = 30$

56. How many integers are between –15 and 24?

Answer: 38 (integers)
Solution: Since it does not include –15 and 24, the total number of integers is (24 – (–15)) – 1 = 39 – 1 = 38.

57. How many nickels are needed to have the same value as 25 quarters?

Answer: 125 (nickels)
Solution: 1 quarter = 25 cents. 1 nickel = 5 cents.
25 quarters = (25 × 25) cents = 625 cents
625 cents = (625 ÷ 5) nickels = 125 nickels

58. Find the remainder when 1001 is divided by 13.

Answer: 0
Solution: $1001 = 7 \times 11 \times 13$. Thus, 1001 is divisible by 13 and the remainder is 0.

59. How many pints are in 5 gallons?

Answer: 40 (pints)
Solution: 1 gallon = 4 quarts. 1 quart = 2 pints.
5 gallons = (5×4) quarts = 20 quarts = (20×2) pints = 40 pints

60. Find the greatest multiple of 65 that is less than 200.

Answer: 195
Solution: $200 \div 65 = 3$ R 5. Thus, $65 \times 3 = 195$ is the greatest multiple of 65 less than 200.

61. Find the value of $4 \times 3 \times 2 \times 1 \times 2 \times 3 \times 4$.

Answer: 576
Solution: $(4 \times 3 \times 2) \times 1 \times (2 \times 3 \times 4) = 24 \times 1 \times 24 = 24^2 = 576$

62. It takes Julia 45 minutes to drive to work. If work starts at 8:11 AM, what time should she leave her house in order to arrive on time?

Answer: 7:26 AM
Solution: 45 minutes before 8:11 AM is 7:26 AM.
Adding 15 minutes and then subtracting an hour is another way to find 7:26 AM.

63. Two dice are rolled. What is the probability that the sum of the two numbers rolled is 14? Express your answer in the simplest form.

Answer: 0
Solution: The largest sum possible is $(6 + 6) = 12$. A sum of 14 is impossible. Therefore, the probability is 0.

64. Find the circumference of a circle with radius 5. Round your answer to the nearest whole number.

Answer: 31
Solution: $C = 2\pi r = 2\pi \times 5 = 10\pi$. π is approximately equal to 3.14.
Thus, $10\pi = 10 \times 3.14 = 31.4$. Rounded to the nearest whole number, the circumference of the circle is 31.

65. Find the range of the set: {23, 54, 0, 11, 32, 72, 16, 100, 88, 96}

Answer: 100
Solution: The smallest number is 0 and the largest number is 100.
The range is $100 - 0 = 100$.

66. 6 fair coins are flipped. What is the probability that 3 of the coins are heads and 4 of the coins are tails? Express your answer in simplest form.

Answer: 0
Solution: Since there are 6 coins, it is impossible to have 3 heads and 4 tails ($3 + 4 = 7$).
Thus, the probability is 0.

67. What is the product of the Least Common Multiple (LCM) and the Greatest Common Factor (GCF) of 8 and 12?

Answer: 96
Solution: The product of the LCM and the GCF is equal to the product of the two original numbers. $8 \times 12 = 96$.

68. Find the area of an isosceles right triangle with one leg equal to 12.

Answer: 72
Solution: An isosceles right triangle has two equal legs. Thus, the area of the triangle is
$$\frac{l^2}{2} = \frac{12^2}{2} = \frac{144}{2} = 72$$

69. What is the remainder when (11 × 9 × 7 × 5 × 3) is divided by 2?

Answer: 1
Solution: The product (11 × 9 × 7 × 5 × 3) is the product of 5 odd integers. Hence, the product will be odd. Therefore, the remainder will be 1 when divided by 2.

70. Find the value of 5! + 4! + 3! + 2! + 1!

Answer: 153
Solution: 5! + 4! + 3! + 2! + 1! = 120 + 24 + 6 + 2 + 1 = 153

71. How many ways can the letters in the word "DARE" be arranged?

Answer: 24 (ways)
Solution: There are 4 different letters. Thus, there are 4! = 24 ways to arrange them.

72. What is the product of the first 10 whole numbers?

Answer: 0
Solution: The first whole number is 0. Thus, the product will be 0.

73. Find the value of 11 + 13 + 15 + 17 + 19.

Answer: 75
Solution: $S_n = \frac{n(a_1 + a_n)}{2} = \frac{5(11+19)}{2} = \frac{5 * 30}{2} = 75$

74. What is the sum of all the angles in a pentagon? Express your answer in degrees.

Answer: 540°
Solution: A pentagon has 5 sides. Thus, the sum of the angles is
$(n-2)180 = (5-2)180 = 3 * 180 = 540 \text{ degrees}$

75. How many miles are in 36,960 feet?

Answer: 7 (miles)
Solution: 1 mile = 5280 feet.
36,960 feet = (36,960 ÷ 5280) miles = 7 miles.

76. Talia watches five 90-minute movies in rapid succession. If she starts at 12:00 PM, what time will she finish?

Answer: 7:30 PM
Solution: $(5 \times 90) = 450$ minutes. $450 \div 60 = 7$ R 30.
Thus, Talia watches movies for 7 hours and 30 minutes. If she starts at 12:00 PM, she will finish at 7:30 PM.

77. What is the perimeter of a square with area 400?

Answer: 80
Solution: $A = s^2 = 400$. $s = 20$. (Note: $20^2 = 400$)
$P = 4s = 4 \times 20 = 80$.

78. What is the remainder when 20^{15} is divided by 100?

Answer: 0
Solution: $20^2 = 400$. 20^{15} will end with two zeros in the tens and the units digit.
Therefore, the remainder when 20^{15} is divided by 100 is 0.

79. Find the mode of the set: {11, 11, 12, 17, 13, 11, 12, 19, 19, 17, 12, 11}.

Answer: 11
Solution: 11 appears 4 times which is more times than all of the other numbers. Thus, 11 is the mode.

80. Find the number of integers from –2015 to 2015, inclusive.

Answer: 4031 (integers)
Solution: $(2015 - (-2015)) + 1 = 4030 + 1 = 4031$

2016 Answer Keys and Solutions

2016 Individual Round Solutions

Answer Key

1. c) 17
2. c) 200
3. d) 96 cups
4. b) 24 units
5. c) 190
6. d) 9 factors
7. d) 15 containers
8. d) 8
9. e) 48 cents
10. d) 9 hrs, 50 min
11. b) 13 days
12. d) 144 units^2
13. d) 24
14. a) 22 hours
15. c) 18π
16. c) 9 boys
17. b) 20
18. e) 6
19. b) $\frac{1}{12}$
20. b) Bella
21. 987
22. 400
23. 20 (cats)
24. 0
25. 8 (units^3)
26. 23
27. 190 (handshakes)
28. 1023
29. 432 (units^2)
30. 40 (years)

1. John's favorite integer is between 10 and 20, inclusive, and contains the digit 7. What is John's favorite integer?

Answer: c) 17

Solution: The only integer between 10 and 20 that contains the digit 7 is 17.

2. Find the product of the range and the mode of the set:
{0, 100, 23, 2, 64, 23, 44, 65, 78, 12, 12, 2, 2}.

Answer: c) 200

Solution: Arrange the set: {0, 2, 2, 2, 12, 12, 23, 23, 44, 64, 65, 78, 100}
The range is (100 – 0) = 100.
2 appears the most. Thus, 2 is the mode.
The product of the range and the mode is (100 × 2) = 200.

3. How many cups are in 6 gallons of water?

Answer: d) 96 cups

Solution: 1 gallon = 4 quarts. 1 quart = 2 pints. 1 pint = 2 cups.
6 gallons = (6 × 4) quarts = 24 quarts = (24 × 2) pints = 48 pints = (48 × 2) cups = 96 cups.

4. The area of a square is equivalent to the area of a rectangle. If the side lengths of the rectangle are 3 units and 12 units, what is the perimeter of the square?

Answer: b) 24 units

Solution: The area of the rectangle is $A = lw = 12 \times 3 = 36$.
Thus, the area of the square is also 36. $A = s^2 = 36$. $s = 6$ units, because $6^2 = 36$.
The perimeter of the square is $P = 4s = 4 \times 6 = 24$ units.

5. What is the value of 1 × 2 + 3 × 4 + 5 × 6 + 7 × 8 + 9 × 10?

Answer: c) 190

Solution: (1 × 2) + (3 × 4) + (5 × 6) + (7 × 8) + (9 × 10) = 2 + 12 + 30 + 56 + 90 = 190.

6. How many positive factors does 36 have?

Answer: d) 9 factors

Solution: $36 = 2^2 3^2$. Thus, the number of factors is $(2 + 1) \times (2 + 1) = 3 \times 3 = 9$

7. A shipping service company packs pots in boxes, then packs boxes in containers. 6 pots fit in 1 box and 8 boxes fit in 1 container. If there are 720 pots to be shipped, what is the least number of containers that are needed?

Answer: d) 15 containers

Solution: 1 container = 8 boxes. 1 box = 6 pots.
720 pots = (720 ÷ 6) boxes = 120 boxes = (120 ÷ 8) containers = 15 containers

8. What is the product of the smallest prime number and the smallest composite number?

Answer: d) 8

Solution: The smallest prime number is 2 which has 2 factors (1, 2). The smallest composite number is 4 which has 3 factors (1, 2, 4).
Thus, the product is $(2 \times 4) = 8$.

9. Dominic loves to collect pennies, nickels and dimes. If he has 3 of each of these coins, how much money does he have?

Answer: e) 48 cents

Solution: 1 penny = 1 cent. 1 nickel = 5 cents. 1 dime = 10 cents.
3 pennies + 3 nickels + 3 dimes = $(3 \times 1) + (3 \times 5) + (3 \times 10) = 3 + 15 + 30 = 48$ cents.

10. Kyle leaves home for work at 7:30 AM and returns home at 6:30 PM. If it takes him 35 minutes to drive from home to work for each trip, how long does he stay at work? (Assuming he does not leave his workplace until he goes home)

Answer: d) 9 hrs and 50 min

Solution: 7:30 AM to 6:30 PM is 11 hours. If it takes him 35 minutes to drive from home to work, he drives for a total of (35 × 2) = 70 minutes or 1 hour and 10 minutes.
11 hours – (1 hour and 10 minutes) = 11 hours – 1 hour – 10 minutes =
10 hours – 10 minutes = 9 hours and 50 minutes.

11. If a woodchuck could chuck 24 lbs of wood in 6 days, how many days would it take to chuck 52 lbs of wood? (Assume it chucks wood at a constant rate.)

Answer: b) 13 days

Solution: Using the work formula, Work = Rate × Time, the woodchuck's rate can be found:
Rate = Work ÷ Time = (24 lbs) ÷ (6 days) = 4 lbs per day.

Now, the work formula can be used again to find how many days it takes to chuck 52 lbs.
Time = Work ÷ Rate = (52 lbs) ÷ (4 lbs per day) = 13 days.

12. A rectangle and a square have the same perimeter. If the side lengths of the rectangle are 5 units and 19 units, what is the area of the square?

Answer: d) 144 units^2

Solution: The perimeter of the rectangle is $P = 2l + 2w = (2 \times 5) + (2 \times 19) = 10 + 38 = 48$.
The perimeter of the square is also 48: $P = 4s = 48$. $s = 48 \div 4 = 12$.
$A = s^2 = 12^2 = 144$ units^2.

13. What is the Greatest Common Factor (GCF) of 720 and 168?

Answer: d) 24

Solution: $720 = 2^4 3^2 5^1$ and $168 = 2^3 3^1 7^1$. The two numbers share the factor $2^3 3^1 = 24$.
Thus, 24 is the GCF.

14. Dane is driving at a speed of 55 miles per hour. Assuming he drives at a constant rate, how long does it take him to travel 1,210 miles?

Answer: a) 22 hours

Solution: Using Distance = Rate × Time, Time = Distance ÷ Rate.

Time = (1,210 miles) ÷ (55 miles per hour) = 22 hours.

15. A circle has circumference 12π. If Jane cuts the circle in half, what is the area of the half-circle?

Answer: c) 18π

Solution: $C = 2\pi r = 12\pi$. $r = 12 \div 2 = 6$.
$A = \pi r^2 = \pi \times 6^2 = 36\pi$.
Since the circle is cut in half, the area would be $36\pi \div 2 = 18\pi$.

16. The ratio of boys to girls in a classroom is 3:5. If there are 24 total students in the class, how many of the students are boys?

Answer: c) 9 boys

Solution: The ratio totals to (3 + 5) = 8 parts. 24 ÷ 8 = 3. Thus, the ratio is multiplied by a factor of 3 to get the number of boys and girls.
Boys:Girls = 3:5 = (3 × 3):(5 × 3) = 9:15 (where 9 + 15 = 24 total students)
Therefore, there are 9 boys.

17. How many diagonals does a regular octagon have?

Answer: b) 20

Solution: $d = \frac{n(n-3)}{2} = \frac{8(8-3)}{2} = \frac{8*5}{2} = 20$.

18. Joe took a number and multiplied it by 2 when he meant to divide it by 2. Then he took that result and added 2 when he meant to subtract 2, ending up with 34. If Joe did what he meant to do, what number would he have ended up with?

Answer: e) 6

Solution: Working backwards, he ended up with 34. Since he added 2 to his previous number, his previous number must be (34 – 2) = 32. Since he multiplied the original number by 2 to get 32, the original number must be (32 ÷ 2) = 16.
Now, the number that he should've ended up with can be found.
Divide the original number by 2: (16 ÷ 2) = 8
Subtract 2 from the result: (8 – 2) = 6.

19. Josh flips a coin and rolls a 6-sided die. What is the probability that he flips a heads and rolls a 5?

Answer: b) $\frac{1}{12}$

Solution: The probability that he flips a heads is ½ and the probability that he rolls a 5 is ⅙.

By the fundamental theorem of counting, the probability that both happen is $\frac{1}{2} * \frac{1}{6} = \frac{1}{12}$.

20. One of three sisters, Anna, Bella, and Celina, ate the last cookie from the cookie jar. When asked by their mother, each responded differently.
Anna: I didn't eat it!
Bella: Celina ate it!
Celina: Anna ate it!
If Anna is the only one telling the truth, who ate the last cookie from the cookie jar?

Answer: b) Bella

Solution: If Anna is the only one telling the truth, then she is correct in saying that she did not eat the cookie. Bella and Celina must be lying. Bella's false statement implies that Celina did not eat the cookie and Celina's false statement implies that Anna did not eat the cookie. Thus, Anna and Celina did not eat the cookie and since one of the three sisters ate the last cookie, Bella ate the last cookie.

21. Find the largest three digit number with distinct digits.

Answer: 987

Solution: The largest digits that are distinct or different are 7, 8 and 9. The three-digit number 987 is the largest configuration of these 3 digits.

22. What is the positive difference between the numerical value of the volume and the numerical value of the surface area of a cube with side length 10?

Answer: 400

Solution: The volume is $V = s^3 = 10^3 = 1000$ and the surface area is $SA = 6s^2 = 6 \times 10^2 = 600$.
Thus, the positive difference is $(1000 - 600) = 400$.

23. The ratio of dogs to cats at an animal shelter was 6:4. After 10 dogs were adopted, the ratio of dogs to cats became 1:1. How many cats are there at the animal shelter?

Answer: 20 (cats)

Solution: The number of cats remained constant. Thus, for every 4 cats, $(6 - 4) = 2$ dogs were adopted in order to give the new ratio $(6 - 2):4 = 4:4$ or 1:1.
$(10 \div 2) = 5$. The ratio is multiplied by a factor of 5.
Dogs:Cats = $6:4 = (6 \times 5):(4 \times 5) = 30:20$ (where there are 10 more dogs than cats).
Thus, there are 20 cats at the animal shelter.

24. Let N be 1,234,567,890,987,654,321. Now, let the product of the digits in N be P and the sum of the digits in N be S. What is the product of P and S?

Answer: 0

Solution: The product of the digits in N is 0 because one of the digits is 0. Regardless of what S is, the product of P and S will be 0 because $P = 0$.

25. What is the volume of a cube with surface area 24 units^2?

Answer: 8 (units^3)

Solution: Surface Area $= 6s^2 = 24$. $s^2 = 24 \div 6 = 4$. $s = 2$.
$V = s^3 = 2^3 = 8$ units^3

26. Let $a + b + c = 13$ and $abc = 11$. If a, b, and c are positive whole numbers, find the value of $ab + bc + ac$.

Answer: 23

Solution: Since $abc = 11$ and a, b, and c are positive whole numbers, let $a = 1$, $b = 1$ and $c = 11$. 11 is a prime number and has 2 factors, 11 and 1. This set of a, b, and c works for $a + b + c = 13$ and $abc = 11$ (where $1 + 1 + 11 = 13$ and $1 \times 1 \times 11 = 11$).
Thus, the answer is $ab + bc + ac = (1 \times 1) + (1 \times 11) + (1 \times 11) = 1 + 11 + 11 = 23$.

27. There are 20 people in a room. If each person shakes hands with everyone else exactly once, how many handshakes take place? (Assume a person does not shake hands with him/herself.)

Answer: 190 (handshakes)

Solution: Each person shakes hands with $(20 - 1) = 19$ people. Thus, $20 \times 19 = 380$. However, this number is divided by 2 because the first person shaking hands with the second person is the same as the second person shaking hands with the first person. Thus, there are $380 \div 2 = 190$ handshakes.

28. Calculate the sum: $1 + 2 + 4 + 8 + 16 + 32 + 64 + 128 + 256 + 512$

Answer: 1023

Solution: $(1 + 2) + 4 + 8 + 16 + 32 + 64 + 128 + 256 + 512 = (3 + 4) + 8 + 16 + 32 + 64 + 128 + 256 + 512 = (7 + 8) + 16 + 32 + 64 + 128 + 256 + 512 = (15 + 16) + 32 + 64 + 128 + 256 + 512$. The pattern is the sum of all the previous terms is equivalent to one less than the next term.
Thus, $(1 + 2 + 4 + 8 + 16 + 32 + 64 + 128 + 256) + 512 = 511 + 512 = 1023$.

29. Find the area of the following pinwheel which is comprised of a square with side length 12 units and 4 isosceles right triangles.

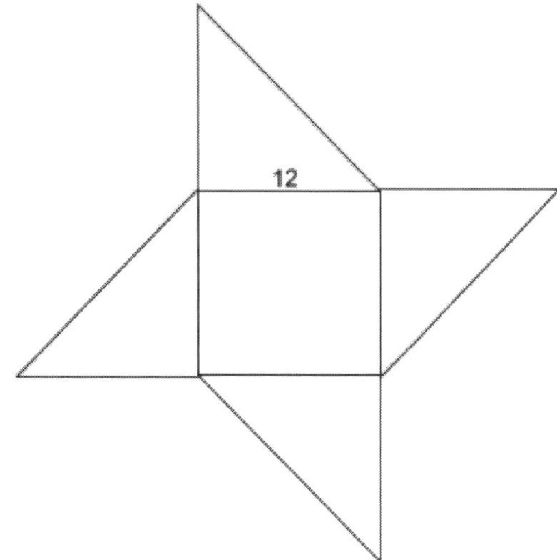

Answer: 432 (units^2)

Solution: There are 4 isosceles right triangles with legs 12 and a square with side length 12.
The area of the square is $A = s^2 = 12^2 = 144$.

The area of one of the isosceles right triangles is $A = \frac{l^2}{2} = \frac{12^2}{2} = 72$

Thus, the total area is Square + 4 Triangles = $144 + (4 \times 72) = 144 + 288 = 432$ units^2.

30. Jackie's dad's age is three times Jackie's age. 8 years ago, her dad's age was eleven times Jackie's age then. What is the sum of Jackie's current age and her dad's current age? Express your answer in years.

Answer: 40 (years)

Solution: The first statement can be written as D = 3J where D is Jackie's dad's age and J is Jackie's age. The second statement can be written as (D – 8) = 11(J – 8).

(1) D = 3J
D – 8 = 11(J – 8)
(2) D – 8 = 11J – 88

Plugging (1) into (2):
3J – 8 = 11J – 88. Subtracting 3J from both sides:
– 8 = 8J – 88. Adding 88 to both sides:
80 = 8J. Dividing both sides by 8:
J = 10. Thus, Jackie's current age is 10.
Since her dad is three times as old as she is, he is D = 3J = 3 × 10 = 30.

The sum of these two is 10 + 30 = 40 years.

2016 Team Round Solutions

Answer Key

1. $\frac{1}{10}$

2. 114 (units)

3. 495

4. 25

5. 66

6. 25 (bicycles)

7. 150 (miles)

8. 108π

9. 9

10. 24 (ways)

1. Given a set: $\{1003, 560, 77, 880, 990099, 2801, 302, 0, 80, 1120\}$. If one of the numbers is chosen at random, what is the probability that the product of all of its digits is positive? Express your answer as a common fraction.

Answer: $\frac{1}{10}$

Solution: There are 10 numbers total where 9 of them have a 0 as a digit. Thus, the product of the digits for the 9 numbers will be 0. The only number that will have a positive product is 77.

Thus, the probability is $\frac{1}{10}$.

2. A regular triangle, a square, a regular pentagon and a regular hexagon all have the same perimeter. If the side length of the pentagon is 24 units, what is the sum of the side length of the triangle, the side length of the square, the side length of the pentagon and the side length of the hexagon?

Answer: 114 (units)

Solution: If the side length of the pentagon is 24, the perimeter is $P_{Pentagon} = 5s = 5 \times 24 = 120$.

Since all the shapes have the same perimeter (120), the side lengths of each can be found:

$s_{Triangle} = 120 \div 3 = 40$ units

$s_{Square} = 120 \div 4 = 30$ units

$s_{Pentagon} = 24$ units

$s_{Hexagon} = 120 \div 6 = 20$ units

The sum of these 4 side lengths is $(40 + 30 + 24 + 20) = 114$ units

3. Find the sum of all the two-digit numbers that are equivalent to the original number when written in reverse order. (For example, 121 is written the same forwards and backwards.)

Answer: 495

Solution: The two-digit numbers that are the same when written in reverse order are: 11, 22, 33, 44, 55, 66, 77, 88 and 99. The sum of these is the sum of an arithmetic sequence because there is a common difference of 11, the first term is 11 and the last term is 99:

$$S_n = \frac{n(a_1 + a_n)}{2} = \frac{9(11 + 99)}{2} = \frac{9 * 110}{2} = 495.$$

4. What is the remainder when 5^{10} is divided by 100?

Answer: 25

Solution: The remainder when 5^{10} is divided by 100 is the same as the last two digits of 5^{10}.
$5^1 = 5$, $5^2 = 25$, $5^3 = 125$, $5^4 = 625$, etc. Every number after 5^1 ends in 25.
Thus, 5^{10} will end in 25.

5. A sequence of numbers follow a rule that, starting with the third term, the next term is the sum of the previous two terms. The sequence is 6, a, b, 12, c, d. If a, b, c, and d are counting numbers, what is the sum of a, b, c, and d?

Answer: 66

Solution: Following the rule, $6 + a = b$ and $a + b = 12$. Thus, a and b can be found using these two equations.
Substituting $6 + a = b$ into $a + b = 12$:
$a + b = a + (6 + a) = 2a + 6 = 12$. Subtracting 6 from both sides:
$2a = 12 - 6 = 6$. Dividing both sides by 2:
$a = 6 \div 2 = 3$. Now, b can be found using $6 + a = b$:
$b = 6 + a = 6 + 3 = 9$.
The sequence becomes: 6, 3, 9, 12, c, d. Now, c and d can be found:
$c = 9 + 12 = 21$
$d = 12 + c = 12 + 21 = 33$.

Therefore, the sum of a, b, c, and d is $(3 + 9 + 21 + 33) = 66$.

6. Johnny owns a store that sells tricycles and bicycles for all ages. In his inventory right now, he has 12 more tricycles than bicycles and there are 161 wheels in total. How many bicycles does Johnny have in his inventory? (Assume tricycles have 3 wheels and bicycles have 2 wheels)

Answer: 25 (bicycles)

Solution: Johnny has 12 more tricycles than bicycles. If he takes out $12 \times 3 = 36$ wheels, he would have the same number of tricycles as bicycles.
The total number of wheels would be $161 - 36 = 125$ wheels.

If the number of tricycles is the same as the number of bicycles, for each pair, there are 2 + 3 = 5 wheels. This can be divided into 125 to find the number of bicycles and tricycles. 125 ÷ 5 = 25.
Thus, there are 25 bicycles and 25 tricycles if the 12 tricycles were taken out. Originally, there are 25 bicycles and 37 tricycles which total to 161 wheels.

7. Two cars, 225 miles apart, are driving towards each other. If the first car is going twice as fast as the second car, how many miles will the first car travel when they meet? (Assume the cars are moving at a constant rate.)

Answer: 150 (miles)

Solution: If the first car goes twice as fast as the second car, it will cover twice the distance as the second car. Speed/Distance Ratio of the 1st Car to 2nd Car is 2:1.
Thus, the first car will travel $\frac{2}{2+1}$ or ⅔ of the total distance. $\frac{2}{3} * 225 = 150$ miles.

8. In the figure below, a rectangle holds 12 congruent circles that are tangent to the rectangle's sides and to each other. If the perimeter of the rectangle is 84, what is the total area of the 12 circles? Express your answer in terms of π.

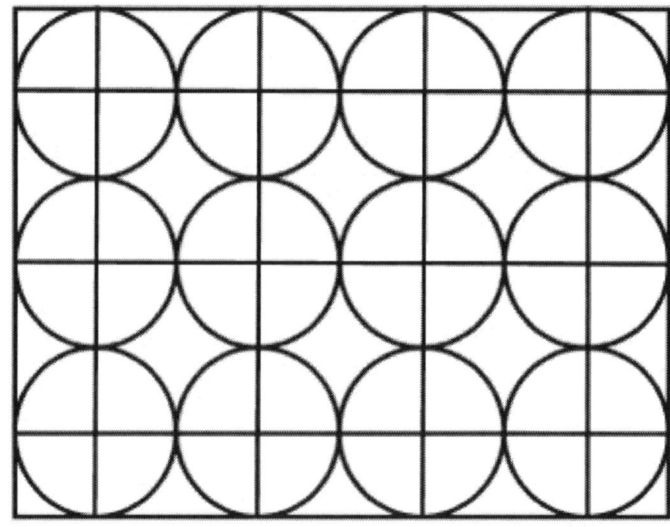

Answer: 108π

Solution: Drawing 7 lines (see figure), the width of the rectangle is equivalent to 6 times the radius of a circle and the length of the rectangle is equivalent to 8 times the radius of a circle.

Thus, W = 6r and L = 8r.
The perimeter of the rectangle is $P = 2L + 2W = (2 \times 8r) + (2 \times 6r) = 16r + 12r = 28r = 84$
Dividing both sides by 28: $r = 84 \div 28 = 3$.
Now the area of one of the circles can be found: $A = \pi r^2 = \pi \times 3^2 = 9\pi$.
Since there are 12 circles, the total area of the 12 circles is $(9\pi \times 12) = 108\pi$.

9. A list of 5 positive integers is written in numerical order from the least to the greatest. If the median, mode, and mean is 4, what is the greatest possible value of the largest number in the list?

Answer: 9

Solution: There are 5 numbers in ascending order: __, __, __, __, __
If the median is 4, then the middle number must be 4: __, __, 4, __, __
To have the greatest possible integer, the first number must be 1: 1, __, 4, __, __
Since 4 is also the mode, the second number should be 2 and the fourth number should be 4 because if 4 was the second number, it would limit the greatest possible integer:

1, 2, 4, 4, __

The final number can be found using the fact that the mean is 4, thus the sum of the 5 integers is $(4 \times 5) = 20$. Subtracting 1, 2, 4 and 4 from 20 will give the largest possible integer:
$20 - (1 + 2 + 4 + 4) = 20 - 11 = 9$. (1, 2, 4, 4, 9)

10. On the top row of a bookshelf, there are 2 math books (Algebra and Geometry) and 3 science books (Biology, Chemistry and Physics). Blake wants to arrange the books in a way where the 2 math books are grouped together and the 3 science books are grouped together. Given that the books can be arranged in their group, how many ways can Blake arrange these 5 books from left to right? (An example is Biology-Chemistry-Physics-Geometry-Algebra.)

Answer: 24 (ways)

Solution: The 2 math books can be arranged $2! = 2$ ways and the 3 science books can be arranged $3! = 6$ ways. In addition, the order can be math books then science books or science books then math books which gives 2 ways of arranging the groups.
By the Fundamental Theorem of Counting, there are $(2 \times 6 \times 2) = 24$ ways.

2016 Countdown Round Solutions

1. 11 (integers)
2. 24
3. 0
4. 104 (pints)
5. 897
6. ⅙
7. 137
8. 121π
9. 132 (cents)
10. 13 (hours)
11. 17 (yards)
12. 325
13. 60 (units)
14. 4
15. 176 (legs)
16. Monday
17. ⅛
18. 209
19. 8 (factors)
20. $12.00
21. 105π
22. 8:16 PM
23. 19
24. 90 (seconds)
25. 4 (factors)
26. 15
27. 192 (cups)
28. 16
29. 7200 (minutes)
30. 8
31. 2 (diagonals)
32. 14
33. 5
34. 9995
35. 20 (ways)
36. 58 (gumballs)
37. 28,800 (seconds)
38. 60 (units)
39. 105
40. ⅓
41. 92 (wheels)
42. 55 (mph)
43. 6 (pairs)
44. 107 (sides)
45. 14 (integers)
46. 7 (hours)
47. 2 (factors)
48. 90 (inches)
49. 42
50. 6:27 PM
51. 35
52. 168π
53. $2.38
54. 300
55. 9 (socks)
56. 6 (factors)
57. 0
58. 30 (teaspoons)
59. 256 (units^2)
60. 6
61. 99
62. $\frac{1}{36}$
63. 20
64. 25
65. 238
66. 20 (multiples)
67. 27 (diagonals)
68. 100 (numbers)
69. 700
70. 120°
71. 55
72. 25,200 (claps)
73. 2002
74. 63 (quarters)
75. 30 (units)
76. 31
77. 4250 (pounds)
78. 4 (minutes)
79. 4 (marbles)
80. 148

81. 6 (ways)
82. 990 (numbers)
83. 16
84. 14
85. 90 (pages)
86. 2
87. 37 (nickels)
88. 288
89. 2500π
90. 3
91. ⅓
92. 80 (words per minute)
93. 300 (toy cars)
94. 25
95. 136 (pears)
96. 120 (combinations)
97. 0
98. 45 (minutes)
99. 20 (percent)
100. 400
101. 72 (toy cars)
102. 440π
103. 64
104. 91 (multiples)
105. 8 (days)
106. 10
107. 25
108. 108 (legs)
109. $\frac{1}{12}$
110. Monday

1. How many integers are between 2 and 12, inclusive?

Answer: 11 (integers)
Solution: (12 – 2) + 1 = 11 integers

2. What is the product of the number of sides in a rectangle and the number of sides in a hexagon?

Answer: 24
Solution: A rectangle has 4 sides and a hexagon has 6 sides. Thus, 4 × 6 = 24.

3. Find the difference between the two sums, (3 + 5 + 7 + 9 + 11) and (11 + 9 + 7 + 5 + 3).

Answer: 0
Solution: The two sums are identical where the numbers are reversed. (3 + 5 + 7 + 9 + 11) is the same as (11 + 9 + 7 + 5 + 3). Thus, the difference is 0.

4. How many pints are in 13 gallons?

Answer: 104 (pints)
Solution: 1 gallon = 4 quarts. 1 quart = 2 pints.
13 gallons = (13 × 4) quarts = 52 quarts = (52 × 2) pints = 104 pints

5. If A = 199, B = 299 and C = 399, what is the value of C + A + B?

Answer: 897
Solution: Let A = 200 – 1, B = 300 – 1, and C = 400 – 1.
Thus, C + A + B = (400 – 1) + (300 – 1) + (200 – 1) = (200 + 300 + 400) – (1 + 1 + 1) = 900 – 3 = 897.

6. A die is rolled. What is the probability that it does not roll a composite or a prime number? Express your answer as a common fraction.

Answer: $\frac{1}{6}$
Solution: The 6 possibilities are {1, 2, 3, 4, 5, 6}. Every number is a prime or composite number except for the number 1. Thus, the probability that it is not a prime or composite is $\frac{1}{6}$.

7. Which of these numbers is a prime number? {112, 137, 184, 156}.

Answer: 137
Solution: The only even prime number is 2. Thus, any other even number must be composite. Since 137 has two factors, 1 and 137, it is a prime number.

8. What is the area of a circle with circumference 22π? Express your answer in terms of π.

Answer: 121π
Solution: $C = 2\pi r = 22\pi$. $r = 22 \div 2 = 11$. Thus, the area is $A = \pi r^2 = \pi \times 11^2 = 121\pi$.

9. Eric has 7 pennies, 4 nickels, 3 dimes and 3 quarters. How many cents does he have?

Answer: 132 cents
Solution: 1 penny = 1 cent. 1 nickel = 5 cents. 1 dime = 10 cents. 1 quarter = 25 cents.
7 pennies + 4 nickels + 3 dimes + 3 quarters = (7 × 1) + (4 × 5) + (3 × 10) + (3 × 25) = 7 + 20 + 30 + 75 = 132 cents.

10. Miranda is driving her new car at a constant speed of 31 miles per hour. How long does it take for her to travel 403 miles? Express your answer in hours.

Answer: 13 (hours)
Solution: Using Distance = Rate × Time,
Time = Distance ÷ Rate = (403 miles) ÷ (31 miles per hour) = 13 hours

11. How many yards are in 612 inches?

Answer: 17 (yards)
Solution: 1 yard = 3 feet. 1 foot = 12 inches.
612 inches = (612 ÷ 12) feet = 51 feet = (51 ÷ 3) yards = 17 yards

12. Find the value of 25 + 45 + 65 + 85 + 105.

Answer: 325
Solution: This is an arithmetic sequence with a common difference of 20.

$$S_n = \frac{n(a_1 + a_n)}{2} = \frac{5(25 + 105)}{2} = \frac{5 * 130}{2} = 5 * 65 = 325$$

13. How much greater is the perimeter of a square with side length 37 units than the perimeter of a square with side length 22 units?

Answer: 60 (units)
Solution: The difference in side length between the two squares is $(37 - 22) = 15$ units. Since there are four sides in a square, the two perimeters differ by $(15 \times 4) = 60$ units.

14. Find the remainder when 123,456,789 is divided by 5.

Answer: 4
Solution: The remainder when divided by 5 is the remainder when the units digit is divided by 5. Thus, the remainder is found by $9 \div 5 = 1$ R 4.

15. Pablo has 34 chickens and 27 horses on his farm. Assuming chickens have 2 legs and horses have 4 legs, how many animal legs are there in total on the farm?

Answer: 176 (legs)
Solution: $(34 \times 2) + (27 \times 4) = 68 + 108 = 176$ legs.

16. John's birthday is on February 2, which is a Tuesday in 2016. If his younger sister's birthday is on February 29, what day of the week is her birthday in 2016?

Answer: Monday
Solution: His younger sister's birthday is 27 days after his birthday. 27 days is 1 day less than 28 days or 4 weeks. Thus, her birthday will be 1 day before a Tuesday (because 4 weeks after Tuesday is a Tuesday). Her birthday will be on a Monday.

17. If three coins are flipped, what is the probability that the outcome is 0 heads?

Answer: $\frac{1}{8}$
Solution: There are $2^3 = 8$ possible outcomes. Only one outcome has 0 heads: Tails-Tails-Tails. Thus, the probability is $\frac{1}{8}$.

18. If $a + b = 30$ and $a = 11$, what is the value of $a \times b$?

Answer: 209
Solution: a can be substituted to solve for b: $a + b = 11 + b = 30$. Subtracting 11 from both sides: $b = 30 - 11 = 19$.
Thus, $a \times b = 11 \times 19 = 209$.

19. How many positive factors does 30 have?

Answer: 8 (factors)
Solution: $30 = 2^1 3^1 5^1$. The number of positive factors is $(1 + 1) \times (1 + 1) \times (1 + 1) = 2 \times 2 \times 2 = 8$.

20. Gabi has $100 to spend at the mall. She first bought a doll for $15 then she saw a movie with popcorn for $27 and lastly she bought a dress for $46. How much money does she have left? Express your answer in dollars.

Answer: $12.00
Solution: $100 – ($15 + $27 + $46) = $100 – ($88) = $12.00

21. Find the positive difference between the area of a circle with radius 13 and the area of a circle with radius 8. Express your answer in terms of π.

Answer: 105π
Solution: The area of the first circle is $A = \pi r^2 = \pi \times 13^2 = 169\pi$ and the area of the second circle is $A = \pi r^2 = \pi \times 8^2 = 64\pi$. The positive difference is $(169\pi - 64\pi) = 105\pi$

22. A theater is screening a documentary that lasts 87 minutes. If the documentary starts at 6:49 PM, what time will it finish?

Answer: 8:16 PM
Solution: 87 minutes is equivalent to 1 hour and 27 minutes. 27 minutes after 6:49 PM is 7:16 PM. 1 hour after 7:16 PM is 8:16 PM.

23. Find the mean of the set: {3, 11, 19, 27, 35}.

Answer: 19
Solution: The set is an arithmetic sequence with a common difference of 8. Thus, the median, which is 19, is also the mean.

24. David is typing at a speed of 74 words per minute. How long does it take him to type 111 words? Express your answer in seconds.

Answer: 90 (seconds)
Solution: $\dfrac{111 \text{ words}}{74 \text{ words per minute}} * \dfrac{60 \text{ seconds}}{1 \text{ minute}} = \dfrac{3}{2} * 60 = 90 \text{ seconds}$.

25. How many positive factors does the number 91 have?

Answer: 4 (factors)
Solution: $91 = 7^1 13^1$. The number of positive factors is $(1 + 1) \times (1 + 1) = 2 \times 2 = 4$.

26. Find the value of $5^2 - 4^2 + 3^2 - 2^2 + 1^2$.

Answer: 15
Solution: $5^2 - 4^2 + 3^2 - 2^2 + 1^2 = 25 - 16 + 9 - 4 + 1 = 15$

27. Kaitlyn needs to buy 12 gallons of water at the store. How many cups of water does she need?

Answer: 192 (cups)
Solution: 1 gallon = 4 quarts. 1 quart = 2 pints. 1 pint = 2 cups.
12 gallons = (12×4) quarts = 48 quarts = (48×2) pints = 96 pints = (96×2) cups = 192 cups

28. What is the Greatest Common Factor (GCF) of 64 and 48?

Answer: 16
Solution: $64 = 2^6$ and $48 = 2^4 3^1$. Thus, the two numbers share $2^4 = 16$ as the GCF.

29. How many minutes are there in five days?

Answer: 7200 (minutes)
Solution: 1 day = 24 hours. 1 hour = 60 minutes.
5 days = (5×24) hours = 120 hours = (120×60) minutes = 7200 minutes

30. Find the remainder when 174,357,818 is divided by 9.

Answer: 8
Solution: The remainder when divided by 9 is equivalent to the remainder when the sum of the digits is divided by 9. The sum of the digits is $(1 + 7 + 4 + 3 + 5 + 7 + 8 + 1 + 8) = 44$.
$44 \div 9 = 4$ R 8.

31. How many diagonals does a rectangle have?

Answer: 2 (diagonals)

Solution: A rectangle has 4 sides. $d = \frac{n(n-3)}{2} = \frac{4(4-3)}{2} = 2$

32. Lauren is adding all the integers from –13 to 14, inclusive. What is her sum?

Answer: 14

Solution: The sum from –13 to 13 is equal to 0 (e.g. –13 + 13 = 0 and –12 + 12 = 0). Thus, the sum will be 14.

33. What is 5% of 20% of 500?

Answer: 5

Solution: $\frac{5}{100} * \frac{20}{100} * 500 = \frac{1}{20} * \frac{1}{5} * 500 = 5$

34. What is the largest four-digit number divisible by 5?

Answer: 9995

Solution: The largest four-digit number that is divisible by 5 is 9995 because it ends in a 5.

35. How many ways can the letters in the name "ALANA" be arranged?

Answer: 20 (ways)

Solution: There are 5 letters and 5! ways to arrange them. However, there are 3 A's which can be arranged 3! ways. Therefore, the total number of ways is $\frac{5!}{3!} = 5 * 4 = 20$.

36. Kady has 234 quarters. If a gumball costs $1, how many gumballs can she buy?

Answer: 58 (gumballs)

Solution: 1 dollar = 4 quarters. 234 ÷ 4 = 58 R 2. Thus, she can buy 58 gumballs.

37. How many seconds are in 8 hours?

Answer: 28,800 (seconds)
Solution: 1 hour = 60 minutes. 1 minute = 60 seconds.
8 hours = (8 × 60) minutes = 480 minutes = (480 × 60) seconds = 28,800 seconds

38. If the area of a square is 225 units^2, what is its perimeter?

Answer: 60 (units)
Solution: $A = s^2 = 225$. Thus, s = 15 because $15^2 = 225$.
P = 4s = 4 × 15 = 60 units

39. What is the sum of all the multiples of 7 between 1 and 40?

Answer: 105
Solution: There are 5 multiples of 7 between 1 and 40 (40 ÷ 7 = 5 R 5). The first multiple is (1 × 7) = 7 and the last multiple is (5 × 7) = 35. The sum of the 5 multiples is the sum of an arithmetic sequence: $S_n = \frac{n(a_1 + a_n)}{2} = \frac{5(7+35)}{2} = \frac{5 * 42}{2} = 105$

40. A bag contains 5 red marbles, 7 blue marbles and 9 green marbles. If one marble is chosen at random out of the bag, what is the probability that it is a blue marble? Express your answer as a common fraction.

Answer: ⅓
Solution: There are (5 + 7 + 9) = 21 marbles total. If there are 7 blue marbles, the probability is $\frac{7}{21}$ or ⅓. $\frac{7}{21}$ or ⅓.

41. There are 17 cars and 12 motorcycles in a parking lot. How many wheels are there in total?

Answer: 92 (wheels)
Solution: A car has 4 wheels and a motorcycle has 2 wheels.
(17 × 4) + (12 × 2) = 68 + 24 = 92 wheels.

42. McKenna drove for 15 hours and traveled 825 miles. On average, how fast was she driving in miles per hour?

Answer: 55 (miles per hour)
Solution: Using Distance = Rate × Time,
Rate = Distance ÷ Time = (825 miles) ÷ (15 hours) = 55 miles per hour.

43. A high school's student council has 4 members: Aisha, Braden, Chiemi and Darla. For a fundraiser, they need 2 members of student council to run it. How many different pairs of members can be selected to run the fundraiser?

Answer: 6 (pairs)
Solution: Each person has 4 – 1 = 3 people that he or she could pair up with.
Thus, 4 × 3 = 12. However, this number is divided by 2 because the pair of Aisha and Braden is the same as the pair of Braden and Aisha. 12 ÷ 2 = 6 pairs.

44. Find the positive difference between the total number of sides of 33 pentagons and the total number of sides of 34 octagons.

Answer: 107 (sides)
Solution: A pentagon has 5 sides and an octagon has 8 sides.
The 33 pentagons have a total of (33 × 5) = 165 sides and the 34 octagons have a total of (34 × 8) = 272 sides
272 – 165 = 107 sides.

45. How many integers between 1 and 50, inclusive, contain the digit 1?

Answer: 14 (integers)
Solution: 5 integers have 1 as a units digit (1, 11, ... , 41) and 10 integers have 1 as a tens digit (10, 11, ... , 19). Since one of these numbers (11) is in both categories, the total number of integers is (5 + 10 – 1) = 14 integers.

46. Poe is flying in his cruiser at a speed of 1370 miles per hour. How many hours will it take him to travel 9590 miles?

Answer: 7 (hours)
Solution: Using Distance = Rate × Time,
Time = Distance ÷ Rate = (9590 miles) ÷ (1370 miles per hour) = 7 hours

47. How many positive factors does the number 101 have?

Answer: 2 (factors)
Solution: 101 is a prime number. Thus, it has 2 factors: 1 and itself.

48. A basketball player's height is 7 feet and 6 inches. What is his height in inches?

Answer: 90 (inches)
Solution: 1 foot = 12 inches.
7 feet + 6 inches = $(7 \times 12) + 6 = 84 + 6 = 90$ inches.

49. Find the median of this set: $\{-113, -99, 16, 68, 1234, 2016\}$.

Answer: 42
Solution: The numbers are already in order. There is an even number of terms, thus, the median is the average of the two middle numbers: $(16 + 68) \div 2 = 84 \div 2 = 42$.

50. Melanie took a break from work for 113 minutes. If she returned to work at 4:34 PM, what time did her break start?

Answer: 6:27 PM
Solution: 113 minutes is 7 minutes less than 2 hours. 2 hours after 4:34 PM is 6:34 PM. 7 minutes before 6:34 PM is 6:27 PM.

51. Steven is thinking of a whole number greater than 30 but less than 40 that is divisible by 5. What number is Steven thinking of?

Answer: 35
Solution: The only whole number that is divisible by 5 between 30 and 40 is 35 which ends in 5. Thus, Steven is thinking of 35.

52. Find the positive difference between the numerical value of the circumference of a circle and the numerical value of the area of a circle with radius 14. Express your answer in terms of π.

Answer: 168π
Solution: $C = 2\pi r = 2\pi \times 14 = 28\pi$. $A = \pi r^2 = \pi \times 14^2 = 196\pi$.
$A - C = 196\pi - 28\pi = 168\pi$.

53. Carver is going to the movies and brings \$18.79. If the movie ticket costs \$10.48 and popcorn costs \$5.93, how much money will he have left after paying for these two items? Express your answer in dollars.

Answer: \$2.38
Solution: \$18.79 – (\$10.48 + \$5.93) = \$18.79 – \$16.41 = \$2.38

54. Gauss loves to add numbers. If he adds all the integers from 1 to 24, inclusive, what sum will he have?

Answer: 300
Solution: $S_n = \frac{n(a_1 + a_n)}{2} = \frac{24(1+24)}{2} = 300$.

55. Josh has a sock drawer that has 4 black socks, 3 yellow socks and 5 purple socks. What is the least number of socks that he must take out to guarantee that he has drawn at least 1 black sock?

Answer: 9 (socks)
Solution: He must draw out the number of non-black socks in addition to one black sock: Yellow + Purple + 1 (Black) = 3 + 5 + 1 = 9 socks.

56. How many positive factors does the number 28 have?

Answer: 6 (factors)
Solution: $28 = 2^2 7^1$. The number of factors is (2 + 1) × (1 + 1) = 3 × 2 = 6.

57. If a die is rolled, what is the probability that a multiple of 7 is rolled? Express your answer in simplest form.

Answer: 0
Solution: None of the 6 possibilities, {1, 2, 3, 4, 5, 6}, is a multiple of 7. Thus, the probability is 0.

58. How many teaspoons are in 10 tablespoons?

Answer: 30 (teaspoons)
Solution: 1 tablespoon = 3 teaspoons.
10 tablespoons = (10 × 3) teaspoons = 30 teaspoons.

59. Find the area of a square with perimeter 64 units.

Answer: 256 ($units^2$)
Solution: P = 4s = 64. s = 64 ÷ 4 = 16. A = $s^2 = 16^2 = 256$ $units^2$.

60. Find the units digit of 2016^2.

Answer: 6
Solution: The units digit of 2016^2 is the same as the units digit of $6^2 = 36$, which is 6. Thus, the units digit of 2016^2 is 6.

61. What is the largest multiple of 11 less than 100?

Answer: 99
Solution: 100 ÷ 11 = 9 R 1. Thus, the largest multiple is 11 × 9 = 99.

62. Two dice are rolled. What is the probability that the sum of the numbers rolled is 2? Express your answer as a common fraction.

Answer: $\frac{1}{36}$
Solution: There are $6^2 = 36$ possible outcomes. The only possibility, when the sum of the numbers rolled is 2, is when 1 and 1 are rolled. Thus, the probability is $\frac{1}{36}$.

63. Bethany adds up the first 20 odd natural numbers while Clark adds up the first 20 even natural numbers. What is the positive difference between these two sums?

Answer: 20
Solution: The first even natural number differs from the first odd natural number by 1 (2 – 1 = 1). The second even natural number differs from the second odd natural number by 1 as well (4 – 3 = 1). Thus, the positive difference between the two sums is 20 (20 × 1 = 20).

64. A square's side length and a circle's radius are equivalent in length. If the circumference of the circle is 10π, what is the square's area?

Answer: 25
Solution: C = 2πr = 10π. r = 10 ÷ 2 = 5.
Since r = s, the area of the square is A = $s^2 = 5^2 = 25$.

65. Find the value of $(2^0 + 16) \times (20^1 - 6)$.

Answer: 238

Solution: $(2^0 + 16) \times (20^1 - 6) = (1 + 16) \times (20 - 6) = 17 \times 14 = 238$

66. How many multiples of 100 are between 1 and 2016?

Answer: 20 (multiples)

Solution: 2016 ÷ 100 = 20 R 16. Thus, there are 20 multiples of 100 between 1 and 2016.

67. How many diagonals does a regular nonagon have?

Answer: 27 (diagonals)

Solution: A nonagon has 9 sides. $d = \frac{n(n-3)}{2} = \frac{9(9-3)}{2} = 27$

68. How many whole numbers are between –879 and 100?

Answer: 100 (whole numbers)

Solution: Whole numbers are 0, 1, 2, 3, etc. The whole numbers between –879 and 100 is from 0 to 99. Thus, there are (99 – 0) + 1 = 100 whole numbers.

69. Lucy is trying to figure out Susan's favorite number. If her favorite number is a three-digit positive integer that is divisible by 7 and 100, what is her favorite number?

Answer: 700

Solution: 7 and 100 only share the factor 1. Thus, her favorite number must be divisible by (7 × 100) = 700. Since 700 is the only multiple that is a three-digit positive integer, her favorite number is 700.

70. What is the measure of one of the angles in a regular hexagon? Express your answer in degrees.

Answer: 120°

Solution: A hexagon has 6 sides. The sum of all its angles is $(n-2)180 = (6-2)180 = 4 * 180 = 720$ or 720°. Since a regular hexagon has equal angles, the measure of one of its angles is (720 ÷ 6) = 120°.

71. Find the value of $1^2 + 2^2 + 3^2 + 4^2 + 5^2$.

Answer: 55
Solution: $1^2 + 2^2 + 3^2 + 4^2 + 5^2 = 1 + 4 + 9 + 16 + 25 = 55$

72. Albert can clap 7 times per second. If he continues at this rate, how many times can he clap in an hour?

Answer: 25,200 (claps)
Solution: 1 hour = 60 minutes. 1 minute = 60 seconds.
1 hour = 60 minutes = (60 × 60) seconds = 3600 seconds.
Using Work = Rate × Time,
Work = (7 claps per second) × (3600 seconds) = 25,200 claps.

73. What is the Least Common Multiple (LCM) of 2 and 1001?

Answer: 2002
Solution: Since the two numbers only share the factor 1, the LCM is the product:
2 × 1001 = 2002.

74. How many quarters are needed to exchange for $15.75?

Answer: 63 (quarters)
Solution: 1 quarter = 25 cents. $15.75 = 1575 cents.
(1575 ÷ 25) = 63 quarters

75. A decagon and a heptagon have the same perimeter. If the side length of the decagon is 21 units, what is the side length of the heptagon?

Answer: 30 (units)
Solution: The perimeter of the decagon is P = 10s = 10 × 21 = 210. Since the decagon and the heptagon have the same perimeter, the side length of the heptagon is (210 ÷ 7) = 30 units.

76. Find the sum of the factors of 16.

Answer: 31
Solution: $16 = 2^4$. The sum of the factors is $(2^0 + 2^1 + 2^2 + 2^3 + 2^4) = 1 + 2 + 4 + 8 + 16 = 31$.

77. A car weighs exactly 2.125 tons. How many pounds does this car weigh?

Answer: 4250 (pounds)
Solution: 1 ton = 2000 pounds.
0.125 tons = ⅛ tons.
2000 × ⅛ = 250 pounds
2 tons = (2 × 2000) pounds = 4000 pounds
Thus, 2.125 tons = 4000 pounds + 250 pounds = 4250 pounds

78. Zofia runs at a constant rate of 5 feet per second. How many minutes will it take her to travel 400 yards?

Answer: 4 (minutes)
Solution: $\frac{5 \text{ feet}}{1 \text{ second}} * \frac{60 \text{ seconds}}{1 \text{ minutes}} * \frac{1 \text{ yard}}{3 \text{ feet}} = \frac{100 \text{ yards}}{1 \text{ minute}}$
Using Distance = Rate × Time,
Time = Distance ÷ Rate = (400 yards) ÷ (100 yards per minute) = 4 minutes

79. A bag contains 2 blue marbles and 3 red marbles. What is the least number of marbles that must be pulled out of the bag to guarantee that at least 1 blue marble is drawn?

Answer: 4 (marbles)
Solution: The number of marbles that must be drawn is the number of non-blue marbles and 1 blue marble. Red Marbles + 1 (blue marble) = 3 + 1 = 4 marbles.

80. Find the value of $20^2 + 20 - 16^2 - 16$.

Answer: 148
Solution: $(20^2 - 16^2) + (20 - 16) = (20 - 16)(20 + 16) + 4 = 4 \times 36 + 4 = 144 + 4 = 148$

81. Joshua, Jessica, Jonathan and Jedidiah are sitting in a car. If there are four different seats including the driver's seat and Jonathan must drive the car, how many ways can the four people sit in the car?

Answer: 6 (ways)
Solution: If Jonathan must sit in the driver's seat, there are three seats which can be arranged by the other three people.
Thus, there are 3! = 6 ways that the four people can be seated.

82. How many whole numbers are between 11 and 1000, inclusive?

Answer: 990 (numbers)
Solution: Using the formula: (b – a) + 1
(1000 – 11) + 1 = 1000 – 11 + 1 = 1000 – 10 = 990.

83. Find the side length of a regular nonagon with perimeter 144.

Answer: 16
Solution: A nonagon has 9 sides.
Thus, the side length of the regular nonagon is 144 ÷ 9 = 16.

84. Find the median of the set: {1, 33, 999, 12, 16, 0}.

Answer: 14
Solution: Arrange the set in increasing order: {0, 1, 12, 16, 33, 999}.
Since there is an even number of terms, the median is the average of the two middle terms.

$$\text{Mean} = \frac{12 + 16}{2} = \frac{28}{2} = 14.$$

85. Jacob read a 300-page book over the period of three days. If he read 87 pages on the first day and 123 pages on the second day, how many pages did he read on the third day?

Answer: 90 (pages)
Solution: 300 – 87 – 123 = 300 – (87 + 123) = 300 – 210 = 90.

86. What is the remainder when 375,123,898 is divided by 8?

Answer: 2
Solution: When dividing a number by 8, the remainder is equal to the remainder when the last three digits are divided by 8.
898 ÷ 8 = 112 R 2.
The remainder when 375,123,898 is divided by 8 would also be 2.

87. Robert wants to buy a pack of gum from a vending machine which costs \$1.85. What is the number of nickels that Robert needs to pay for the gum?

Answer: 37 (nickels)
Solution: A nickel is worth 5 cents or \$0.05.
$\$1.85 \div \$0.05 = 185 \div 5 = 37$.

88. Find the value of $(1^1 + 2^2 + 3^3 + 4^4)$.

Answer: 288
Solution: $1^1 = 1$; $2^2 = 4$; $3^3 = 27$; $4^4 = 256$
$1 + 4 + 27 + 256 = 288$.

89. A square's side length is equivalent to the diameter of a circle. If the square's perimeter is 400, what is the area of the circle? Express your answer in terms of π.

Answer: 2500π
Solution: If the square's perimeter is 400, the side length is $400 \div 4 = 100$.
($P = 4s$. $s = P \div 4$.)
From the first statement, the side length is equivalent to the diameter of the circle which is 100. The radius of the circle would be $100 \div 2 = 50$. ($d = 2r$. $r = d \div 2$.)
Now, find the area of the circle: $\pi r^2 = \pi \times 50^2 = 2500\pi$.

90. What is the largest prime factor of 108?

Answer: 3
Solution: Find the prime factorization: $108 = 2 \times 54 = 2^2 \times 27 = 2^2 \times 3 \times 9 = 2^2 3^3$
The largest prime factor of 108 is 3.

91. Consider all the arrangements of the letters in the word "ACE." What fraction of the arrangements start with the letter "A"? Express your answer as a common fraction.

Answer: $\frac{1}{3}$
Solution: Since each letter is equally likely to be the first letter in the arrangement, $\frac{1}{3}$ of the arrangements will start with the letter "A".
Another way is to list all the arrangements: ACE, AEC, CAE, CEA, EAC, ECA. Two out of the six arrangements start with the letter "A" which gives $\frac{2}{6} = \frac{1}{3}$.

92. Chris can type 1200 words in 15 minutes. On average, how many words can he type per minute?

Answer: 80 (words per minute)
Solution: 1200 words ÷ 15 minutes = 80 words per minute.

93. Jake has 20 toy cars. If Julia has three times as many toy cars as Jake and Jack has five times as many toy cars as Julia, how many toy cars does Jack have?

Answer: 300 (toy cars)
Solution: Julia has three times as many toy cars as Jake which means she has 20 × 3 = 60 toy cars. Jake has five times as many toy cars as Julia which means he has 60 × 5 = 300 toy cars.

94. Find the other leg of a right triangle with area 100 and leg 8.

Answer: 25

Solution:
$$\text{Area of a Right Triangle} = \frac{8 * L}{2} = 4L = 100$$
L = 100 ÷ 4 = 25.

95. In a barn, the ratio of apples to pears is 1 to 2. If there are 68 apples in the barn, how many pears are there?

Answer: 136 (pears)
Solution: If the ratio of apples to pears is 1 to 2, then the number of pears is twice the number of apples. Thus, pears = 68 × 2 = 136.

96. Lucy has 12 different colored dresses and 11 different colored bows. How many different dress-bow combinations can Lucy make if she does not want to wear her red bow?

Answer: 120 (combinations)
Solution: Since Lucy does not want to wear her red bow, she has 11 – 1 = 10 different colored bows to choose from.
Using the Fundamental Theorem of Counting, there are 12 × 10 = 120 combinations.

97. What is the product of all the whole numbers between 0 and 6, inclusive?

Answer: 0
Solution: The product of 0 and another whole number will always be 0.
Thus, the product is 0.

98. Joseph flies an airplane at a constant speed of 600 miles per hour. How many minutes will it take him to travel 450 miles?

Answer: 45 (minutes)
Solution: Using D = R × T. T = D ÷ R.
T = 450 miles ÷ (600 miles per hour) = ¾ hour.
1 hour = 60 minutes. ¾ hour = ¾ × 60 minutes = 45 minutes.

99. If an integer is chosen at random between 1 and 10, inclusive, what is the probability that the integer contains the digit 1? Express your answer as a percent.

Answer: 20 (percent)
Solution: The only integers between 1 and 10, inclusive, that contain the digit 1 are the numbers 1 and 10. Thus, 2 out of 10 numbers contain the digit 1, or 20 percent.

100. Find the value of $(101^2 - 99^2)$.

Answer: 400
Solution: Using the difference of two squares:
$101^2 - 99^2 = (101 - 99)(101 + 99) = 2 \times 200 = 400$.

101. Carlos has a one square foot wooden board. If a toy car takes up two square inches of space, what is the least number of toy cars that can fill the wooden board's area?

Answer: 72 (toy cars)
Solution: $1\ ft^2 = 1\ ft \times 1\ ft = 12\ in \times 12\ in = 144\ in^2$
$1\ ft^2 \div 2\ in^2 = 144\ in^2 \div 2\ in^2 = 72$.

102. Find the numerical sum of the circumference and the area of a circle with diameter 40. Express your answer in terms of π.

Answer: 440π
Solution: The diameter is 40 and the radius is $40 \div 2 = 20$. ($d = 2r$. $r = d \div 2$.)
Circumference = $\pi d = 40\pi$
Area = $\pi r^2 = \pi \times 20^2 = 400\pi$
Circumference + Area = $40\pi + 400\pi = 440\pi$

103. Jordan wants to find the cube of a number with her calculator. Instead of finding the cube, she accidentally calculated the perfect square of the positive number and got 16. What number should she have gotten if she found the cube of the number?

Answer: 64
Solution: If she accidentally calculated the perfect square of a positive number, then the original number is $\sqrt{16} = 4$. Thus, the cube of the number is $4^3 = 64$.

104. How many multiples of 7 are between 70 and 700, inclusive?

Answer: 91 (multiples)
Solution: $70 = 7 \times 10$ and $700 = 7 \times 100$.
The multiples of 7 between 70 and 700, inclusive, are: $7 \times 10, 7 \times 11, 7 \times 12, \ldots, 7 \times 100$.
The number of multiples is the same as the numbers between 10 and 100, inclusive.
Using the formula: $(b - a) + 1 = (100 - 10) + 1 = 91$.

105. Adam bought three gallons of milk. If he drinks three pints of milk a day, how many days will the three gallons of milk last?

Answer: 8 (days)
Solution: 1 gallon = 4 quarts. 1 quart = 2 pints.
3 gallons = (3×4) quarts = 12 quarts = (12×2) pints = 24 pints.
24 pints $\div$ 3 pints = 8 days.

106. What is the positive difference between the sum of all the even numbers and the sum of all the odd numbers between 1 and 20, inclusive?

Answer: 10
Solution: There are the first 10 even numbers and there are the first 10 odd numbers between 1 and 20, inclusive.

The respective sums are $10(10 + 1) = 110$ and $10^2 = 100$ (See Formulas and Tips).
$110 - 100 = 10$.
Another way to solve the problem is to notice that for every pair of two numbers such as (1 and 2) or (3 and 4), they differ by 1. Since there are 10 pairs, the positive difference is $10 \times 1 = 10$.

107. A square and a regular pentagon has the same perimeter. If the side length of the pentagon is 4, what is the area of the square?

Answer: 25
Solution: If the side length of the pentagon is 4, the perimeter is $P = 5s = 5 \times 4 = 20$.
The perimeter is the same for the square which means the side length of the square is $20 \div 4 = 5$. ($P = 4s$. $s = P \div 4$.)
The area of the square is $A = s^2 = 5^2 = 25$.

108. A farm has 24 chickens and 15 cows. If chickens have two legs and cows have four legs, how many animal legs are there in total?

Answer: 108 (legs)
Solution: $(24 \times 2) + (15 \times 4) = 48 + 60 = 108$ legs.

109. Gerald rolls two fair dice. What is the probability that the sum of the two numbers rolled is 11 or 12? Express your answer as a common fraction.

Answer: $\frac{1}{12}$
Solution: There are two ways for the sum to be 11 (5 and 6 or 6 and 5) and one way for the sum to be 12 (6 and 6). By the fundamental theorem of counting, there are $6 \times 6 = 36$ total combinations for the two dice.
Thus, the probability is $\frac{2+1}{36} = \frac{3}{36} = \frac{1}{12}$.

110. James was born on July 1 which was a Saturday. His best friend, Justin was born on July 31 the same year as James. What day of the week was Justin's birth date?

Answer: Monday
Solution: If July 1 was a Saturday, then exactly 4 weeks later on July (1 + 28) or July 29, it should also be a Saturday. This means July 30 was a Sunday and July 31 was a Monday.

Formulas, Strategies, and Tips

Strategies and Tips

1) Read the question twice before attempting to answer. This will help you understand what the question is asking.
2) Before choosing or writing down an answer, read what the question is asking for.
3) Underline key words, especially the units.
4) Leave numbers in prime factorization.
5) If a problem seems too hard, skip it and return to it later. If it is hard for you, it is probably hard for someone else too.
6) Keep track of time. (Use a stopwatch or timer.)
7) Make sure to have all your supplies: pencils, erasers and watch or timer.
8) When you have time remaining, check your work. Solving problems in a different way helps to ensure that you got the correct answer.
9) Write neatly so it is easy to follow when checking your work later on.
10) Most importantly, have fun!

Key Words for Word Problems

Equal = Is, Are, Make

Addition = Plus, And, Altogether, More Than, Sum

Subtraction = Minus, Takes Away, How Much More/Less, Difference

Multiplication = Times, Of, Factor, Each, Product

Division = Divided By, Divided Into, Shared Equally, Quotient

Distinct = Different

Respectively

For example: The ratio of cats to dogs to birds is 3:5:7, respectively. This means that in the ratio, cats equal 3 parts, dogs equal 5 parts and birds equal 7 parts out of the whole.

Dozen = 12

Decimal representation = A single number that is represented with only digits

Definitions and Terms

The terms listed below will appear in this book and in math competitions.

The Basic Terms:

Positive Numbers:	Any number greater than 0. {1, 2, 4.5, … }
Zero or 0.	The only real number that is neither positive nor negative.
Negative Numbers:	Any number less than 0. {−1, −3, −4.5, … }

Natural Numbers or $\mathbb{N}$ consist of all the positive numbers excluding fractions and decimals. They are also known as counting numbers or positive integers.
$\mathbb{N}$ | {1, 2, 3, 4, 5, … }

Whole Numbers consist of 0 and the Natural Numbers. They are also known as nonnegative integers.
Whole Numbers | {0, 1, 2, 3, 4, 5, … }

Integers or $\mathbb{Z}$ consist of 0, all positive and negative numbers excluding fractions and decimals.
$\mathbb{Z}$ | { … −5, −4, −3, −2, −1, 0, 1, 2, 3, 4, 5, … }

Rational Numbers or $\mathbb{Q}$ consist of all the numbers that can be expressed as a ratio of two integers. For example, a simple fraction is a rational number. In addition, integers are rational numbers because they can be expressed as a ratio of the integer and 1.
{i.e. ½, 5, $-\frac{13}{8}$}

Irrational Numbers or $\mathbb{I}$ consist of all the numbers that cannot be expressed as a ratio of two integers.
{i.e. π, $\sqrt{2}$}

Real Numbers or $\mathbb{R}$ are the combination of the rational and irrational numbers.

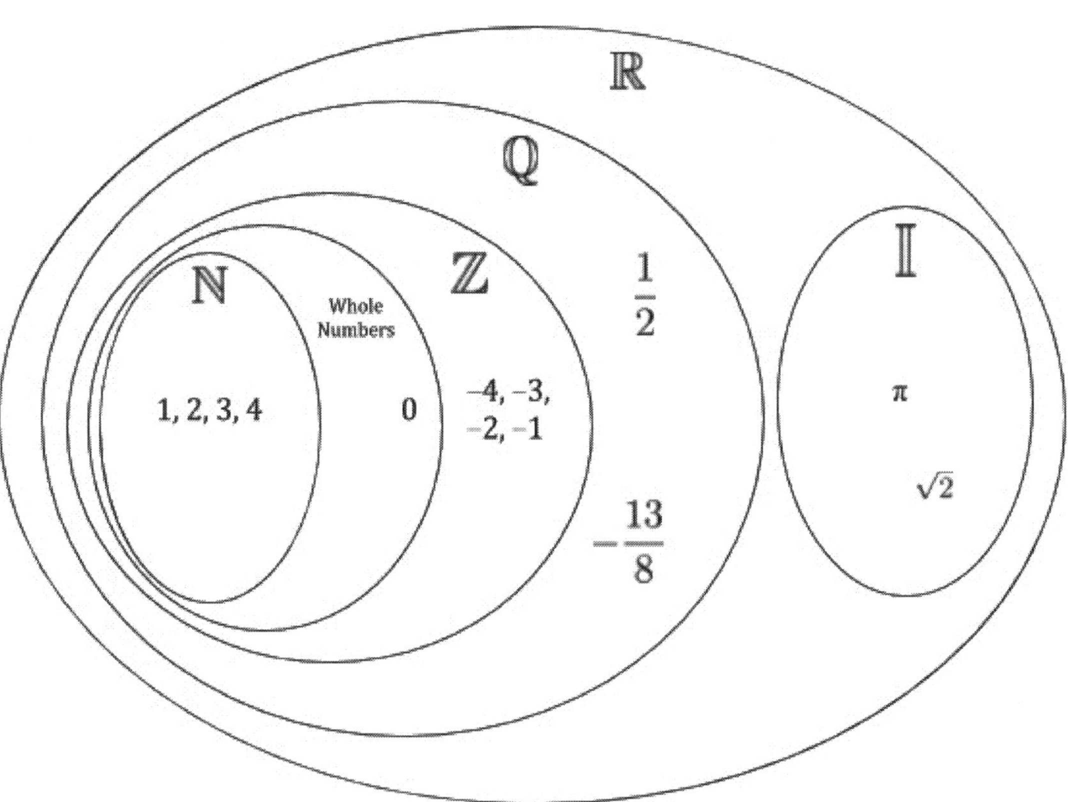

The Four Operations (+, −, ×, ÷)

Addition (+)
$1 + 2 = 3$
1 = augend; 2 = addend; 3 = **sum**

Subtraction (−)
$5 - 4 = 1$
5 = minuend; 4 = subtrahend; 1 = **difference**

Multiplication (× or ∗)
$2 \times 4 = 8$
2 = multiplicand; 4 = multiplier; 8 = **product**

Division (÷ or /)
$7 \div 2 = 3$ R 1
7 = dividend; 2 = divisor; 3 = **quotient**; 1 = **remainder**

(Tip: Bolded math vocabulary words should be memorized.)

Facts to Memorize

Order of Operations (also known as PEMDAS)

The Order of Operations consist of 4 steps:

Step 1) **P**arentheses: ( )
Step 2) **E**xponents: Powers and Square Roots (For example, $5^2 = 5 \times 5 = 25$ in which the exponent, 2, indicates how many times the number is multiplied.)
Step 3) (From left to right)
Multiplication: (×)
Division: (÷)
Step 4) (From left to right)
Addition: (+)
Subtraction: (–)

$7 \times 11 \times 13 = 1001$

Perfect Squares

$1^2 = 1$	$2^2 = 4$	$3^2 = 9$	$4^2 = 16$	$5^2 = 25$
$6^2 = 36$	$7^2 = 49$	$8^2 = 64$	$9^2 = 81$	$10^2 = 100$
$11^2 = 121$	$12^2 = 144$	$13^2 = 169$	$14^2 = 196$	$15^2 = 225$
$16^2 = 256$	$17^2 = 289$	$18^2 = 324$	$19^2 = 361$	$20^2 = 400$
$21^2 = 441$	$22^2 = 484$	$23^2 = 529$	$24^2 = 576$	$25^2 = 625$
$26^2 = 676$	$27^2 = 729$	$28^2 = 784$	$29^2 = 841$	$30^2 = 900$
$31^2 = 961$	$32^2 = 1024$	$33^2 = 1089$	$34^2 = 1156$	$35^2 = 1225$

Perfect Cubes

$1^3 = 1$	$2^3 = 8$	$3^3 = 27$	$4^3 = 64$	$5^3 = 125$
$6^3 = 216$	$7^3 = 343$	$8^3 = 512$	$9^3 = 729$	$10^3 = 1000$

$11^3 = 1331$ $12^3 = 1728$

Powers of Two

$2^1 = 2$	$2^2 = 4$	$2^3 = 8$	$2^4 = 16$	$2^5 = 32$
$2^6 = 64$	$2^7 = 128$	$2^8 = 256$	$2^9 = 512$	$2^{10} = 1024$

First Prime Numbers

2, 3, 5, 7
11, 13, 17, 19
23, 29
31, 37
41, 43, 47
53, 59
61, 67
71, 73, 79
83, 89
97
101, 103, 107, 109

Units of Measurements

Length or Distance Conversions
1 foot = 12 inches
1 yard = 3 feet
1 mile = 5,280 feet
1 meter = 100 centimeters
1 kilometer = 1,000 meters

Time Conversions
1 leap year = 366 days
1 non leap year = 365 days
1 minute = 60 seconds
1 hour = 60 minutes
1 day = 24 hours
1 year = 12 months

Volume Conversions
1 liter = 1,000 cubic centimeters
1 cubic yard = 27 cubic feet
1 cubic foot = 1728 cubic inches

Capacity Conversion
1 cup = 8 fluid ounces
1 pint = 2 cups
1 quart = 2 pints
1 gallon = 4 quarts
1 liter = 1,000 milliliters
1 tablespoon = 3 teaspoons
2 tablespoons = 1 fluid ounce
1 quart ≈ 1 liter

Area Conversion
1 square foot = 144 square inches
1 square yard = 9 square feet
1 square mile = 640 acres

Money
1¢ = $0.01
$1.00 = 100¢

Weight and Mass Conversion
1 pound = 16 ounces
1 kilogram = 1,000 grams
1 ton = 2,000 pounds

Factor Label Method

When converting from one unit to another, the factor label method can be used to keep track of the units. In this method, conversion factors, which can be found above, are set up as fractions that are essentially equal to 1. The conversion factors are multiplied so the units are on the top and on the bottom, thus cancelling units

Follow these steps when using the Factor Label Method

Step 1) Setup the first fraction
Step 2) Line up conversion factors to alternate units
Step 3) Cancel the units
Step 4) Multiply the numbers and keep the remaining unit(s)

Fractions and Percentages

A **fraction** is the ratio of two numbers. It can be written with two numbers separated by a line. The number on the top is called the **numerator** and the number on the bottom is called the **denominator**.

$$\text{Fraction} = \frac{\text{Numerator}}{\text{Denominator}} = \text{Numerator} \div \text{Denominator}$$

Types of Fractions:
A **proper fraction** is a fraction less than 1. An **improper fraction** is a fraction greater than 1. A **mixed number** is a number with a non-zero integer in front and a proper fraction at the end. A **common fraction** is a proper or an improper fraction that is reduced.

Conversion Table (Fraction to Decimal to Percent):

Fraction	Decimal	Percent
0	0	0%
1	1	100%
½	0.50	50%
⅓	$0.\overline{3}$	$33.\overline{3}$%
⅔	$0.\overline{6}$	$66.\overline{6}$%
⅙	$0.1\overline{6}$	$16.\overline{6}$%
⅚	$0.8\overline{3}$	$83.\overline{3}$%
¼	0.25	25%
¾	0.75	75%
⅕	0.2	20%
⅖	0.4	40%
⅗	0.6	60%
⅘	0.8	80%
⅛	0.125	12.5%
⅜	0.375	37.5%
⅝	0.625	62.5%
⅞	0.875	87.5%

Algebra

Counting Integers

When counting the number of integers between two integers *a* and *b*, it can either be inclusive meaning the numbers *a* and *b* are included in the count or not inclusive.

The number of integers between *a* and *b*, inclusive, is $(b-a)+1$.
The number of integers between *a* and *b* is $(b-a)-1$.

Arithmetic Sequences

An **arithmetic sequence** is a sequence of numbers where the difference between consecutive terms is the same value. For example, the following is an arithmetic sequence where 2 is the first term: 2, 6, 10, 14, 18, …

The **common difference** is the difference between any two consecutive terms. In the example above, the common difference would be: 10 – 6 = 4 or 6 – 2 = 4.

The *n*th term of an arithmetic sequence: $a_n = a_1 + (n-1)d$, where a_n is the *n*th term, a_1 is the first term and d is the common difference.
$d = a_2 - a_1 = a_3 - a_2 = \ldots$

The sum of an arithmetic sequence: $S_n = \dfrac{n(a_1 + a_n)}{2}$, where S_n is the sum of the *n* terms, a_1 is the first term of the sequence and a_n is the *n*th term of the sequence.

Sum of the first *n* positive even numbers: $S_n = n(n+1) = n^2 + n$

Sum of the first *n* positive odd numbers: $S_n = n^2$

Distance and Work Formula

Distance = Rate × Time
Work = Rate × Time
Work = People × Rate × Time
(Tip: Make sure the units are consistent when using this formula)

Difference of Two Squares

$a^2 - b^2 = (a - b)(a + b)$
(Tip: This formula is useful for the Pythagorean Theorem)

Fractions to Repeating Decimals

$\frac{a}{9} = 0.aaa...$

$\frac{ab}{99} = 0.abab...$

$\frac{abc}{999} = 0.abcabc...$

etc…

Geometry

Polygons

A **polygon** is a figure with three or more sides. A **regular polygon** has angles and sides that are all equivalent. An **equilateral polygon** has all of its sides equal and an **equiangular polygon** has all of its angles equal.

A **triangle** has three sides.
A **quadrilateral** has four sides.
A **pentagon** has five sides.
A **hexagon** has six sides.
A **heptagon** has seven sides.
An **octagon** has eight sides.
A **nonagon** has nine sides.
A **decagon** has ten sides.

Sum of all the angles in an *n*-gon: Sum = $(n - 2)180$

The number of diagonals in a convex *n*-gon is: $d = \frac{n(n-3)}{2}$ where *n* is the number of sides of the polygon.

Perimeter

The **perimeter** is the sum of all the side lengths.

Triangle: $P = a + b + c$
Equilateral Triangle: $P = 3s$
Rectangle: $P = 2l + 2w$
Quadrilateral: $P = a + b + c + d$
Square: $P = 4s$

Area

The **area** is the amount of space inside a figure.

Triangle: A = $\frac{bh}{2}$

Right Triangle: A = $\frac{l_1 l_2}{2}$

Square: A = s^2

Rectangle: A = lw

Circles

A **circle** is a shape that is equal distance from its center. The distance from its center to any point on the circle is called the **radius** and is denoted by r. The diameter of a circle is equal to twice its radius and is denoted by d.

The **circumference** is the length around the circle.
C = $2\pi r = \pi d$, where π or pi is the ratio of the circumference to the diameter of a circle and is approximately equal to 3.14.

$\pi \approx 3.14$

The area of a circle is: A = πr^2

Pythagorean Theorem

In a right triangle, where one of the angles is 90 degrees, the sum of the squares of the two legs is equivalent to the square of the hypotenuse.
$a^2 + b^2 = c^2$, where a and b are the two legs of the right triangle and c is the hypotenuse which is the side opposite of the right angle.

To find an unknown leg a, the pythagorean theorem can be used:
$a^2 + b^2 = c^2$. $a^2 = c^2 - b^2$. $a = \sqrt{c^2 - b^2}$.

A **pythagorean triple** is a set of three positive integers that makes up the sides of a right triangle.

Pythagorean Triples:

3-4-5	5-12-13	7-24-25
8-15-17	9-40-41	11-60-61
20-21-29		

Using 3-4-5, triples like 6-8-10, 9-12-15, 12-16-20 and 15-20-25 can be made by multiplying the triple (3-4-5) by a factor of 2, 3, 4 and 5, respectively.

3-Dimensional (3D) Figures

A 3D figure consists of vertices, edges and faces. A **face** is a two-dimensional surface that has an area. A **vertex** is a point or a corner where the faces meet. An **edge** is a line segment that connects two vertices and is where the faces meet.

Euler's Formula: V + F – E = 2
The sum of the number of vertices and the number of faces, and subtracting the number of edges of a 3D figure will equal to 2.

A **cube** has six square faces, twelve edges and eight vertices. Its twelve edges are equal in length.

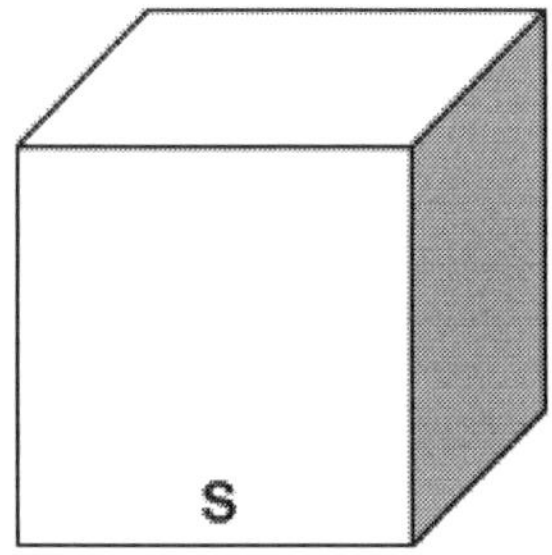

A **rectangular solid** has six rectangular faces, twelve edges, and eight vertices. This figure has a length, a width and a height.

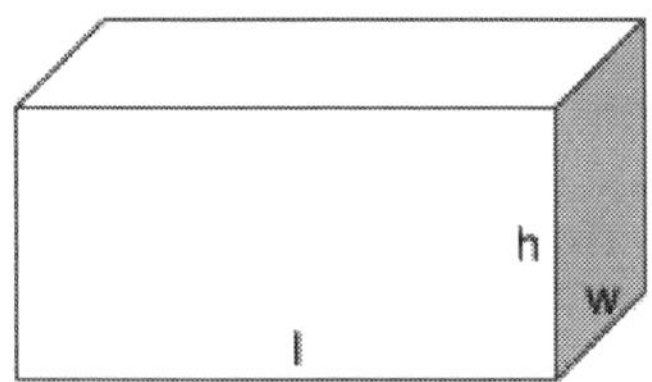

A **sphere** is a solid round ball with a radius.

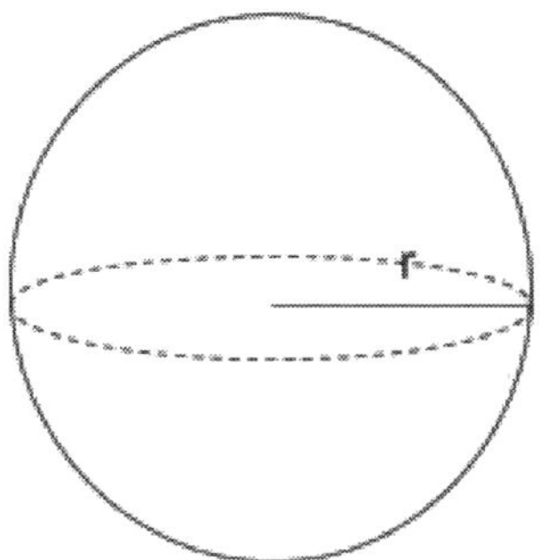

The **volume** of a 3D figure is the amount of space that it occupies. The **surface area** is the sum of the areas of all of its faces.

Volume

Cube: $V = s^3$
Rectangular Solid: $V = lwh$
Sphere: $V = \frac{4\pi r^2}{3}$

Surface Area

Cube: $S_A = 6s^2$
Rectangular Solid: $S_A = 2(lw + wh + lh)$
Sphere: $S_A = 4\pi r^2$

Counting & Probability

Probability is the likeliness or chance that an event is going to occur and is measured by the ratio of favorable outcomes to the total number of possible outcomes. The probability of an event occurring is always between 0 and 1, inclusive.

$$\text{Probability} = \frac{\text{Favorable Outcomes}}{\text{Total Number of Possible Outcomes}}$$

The sum of all the probabilities that an event could or could not occur is always 1.

In other words, $\text{P(Event Occurring)} + \text{P(Event Not Occurring)} = 1$

Sometimes, finding the probability of an event not occurring and subtracting that probability from 1 is more efficient than calculating the probability of an event occurring. This is known as **complementary counting**.

Fundamental Theorem of Counting

If there are x ways of an event occurring and y ways of another event occurring, the number of total possible outcomes is $x \times y$. This is known as the **Fundamental Theorem of Counting**.

Useful Facts:
There are 10 digits (0 to 9).
There are 26 letters in the alphabet (A to Z).
There are 5 vowels in the alphabet (A, E, I, O, U).
There are 21 consonants (non-vowels) in the alphabet (26 – 5 = 21).
There are 2 sides of a coin (head and tail)
There are 6 sides or faces of a die (1 to 6).

Factorials

The factorial of a natural number, *n*, is the product of all the natural numbers less than or equal to *n*. $n! = n * (n-1) * ... * 3 * 2 * 1$.

Proportions and Ratios

Ratio of *X* to *Y* = *X* to *Y* = *X* : *Y* = $\frac{X}{Y}$

If $\frac{A}{B} = \frac{C}{D}$ where $B \neq 0, D \neq 0$, then $AD = BC$.

Odds are used to express the chance of something occurring or not occurring.
Odds for or in favor = Ratio of Number of Successes to Number of Failures
Odds against or not in favor = Ratio of Number of Failures to Number of Successes

Thus, the probability of something occurring is $\frac{\text{Successes}}{\text{Successes} + \text{Failures}}$

Statistics

Given a set of numbers, we can find the mean, median, mode and range.

Mean = Average value = Arithmetic Mean

Mean = $\frac{(\text{Sum of All Terms})}{(\text{Number of Terms})}$

Thus, $(\text{Sum of All Terms}) = (\text{Number of Terms}) * (\text{Mean})$

Median = Middle value
In a set containing an odd number of terms, the median is the middle number.
In a set containing an even number of terms, the median is the average of the two middle numbers.
(Note: For a set that follows an arithmetic sequence, the mean is equal to the median.)

Mode = The value that appears the most
Note: There may be more than one mode. If all of the numbers are different, there is no mode.

Range = The difference between the largest number and the smallest number
Range = Largest Value – Smallest Value

Steps to solve: Mean, Median, Mode and Range Problems
1) Arrange the numbers from least to greatest
2) Count the number of terms
3) Calculate the Mean, Median, Mode and Range

Number Theory

Divisibility Rules

A number is divisible by…
2: if the number is even or ends in 0, 2, 4, 6 or 8
3: if the sum of the digits is divisible by 3
4: if the last two digits make a two-digit number that is divisible by 4
5: if the number ends in 0 or 5
6: if the number is divisible by 2 and 3
8: if the last three digits make a three-digit number that is divisible by 8
9: if the sum of the digits is divisible by 9
10: if the number ends in 0
11: if the difference between the sum of the even-numbered digits and the sum of the odd-numbered digits is 0 or a multiple of 11

A **factor** is a positive integer that can be divided evenly into a number.

A **prime** number is a positive integer that has 2 distinct factors, 1 and itself
A **composite** number is a positive integer that has 3 or more factors.
Note: 1 is not a prime number.

Prime Factorization

The prime factorization of a number is the unique set of prime numbers that, when multiplied, make up the number. For example, the prime factorization of 72 is $2^3 3^2$ where 2 and 3 are prime numbers.

To find the number of factors of a number, take each exponent from the prime factorization and add one to each. Then, calculate the product. For example, the number of factors in 72 is $(3 + 1) \times (2 + 1) = 4 \times 3 = 12$.

To find the sum of factors in a number, find the sum of the powers of each prime factor, and then calculate the product of these sums. For example, the sum of the factors in 72 is $(2^0 + 2^1 + 2^2 + 2^3) \times (3^0 + 3^1 + 3^2) = 15 \times 13 = 195$.

The **Greatest Common Factor** (GCF) of two numbers is the largest positive number that can divided into both numbers.

The **Least Common Multiple** (LCM) of two numbers is the smallest positive number that is divisible by the two numbers.

Odd Number × Odd Number = Odd Number
Odd Number × Even Number = Even Number
Even Number × Even Number = Even Number

About The Authors

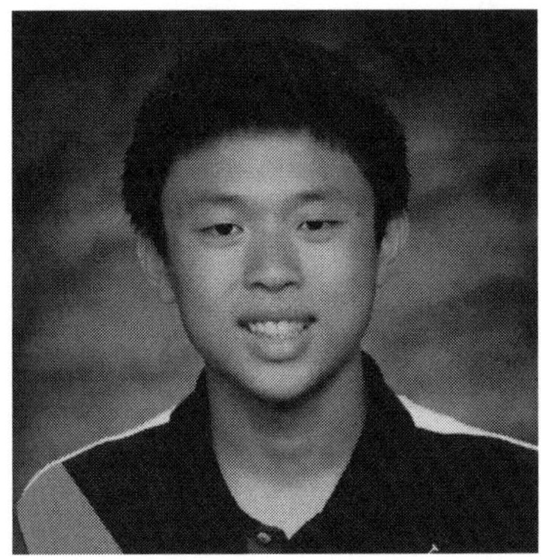

Steven Doan

Steven Doan started participating in math competitions in fifth grade. He earned perfect scores on the math section of the ACT and the PSAT. As a ninth grader, he scored perfect 5s on the AP Calculus BC and AP Statistics exams. Steven is currently a high school sophomore from Houston, Texas.

Below are some of Steven's achievements:

- First Place Individual in Hawaii, American Mathematics Competitions 10 (AMC 10) (2016, 2017)
- American Invitational Mathematical Examination (AIME) Qualifier (2017)
- Co-Captain of the First Place Team, Hawaii State Math Bowl (2017)
- 68th Place Individual in the USA, MATHCOUNTS Nationals (2015)
- MATHCOUNTS Nationals Alumnus (2015, 2016)
- First Place Individual in Hawaii, MATHCOUNTS State Competition (2016)
- First Place Individual in Hawaii, American Mathematics Competitions 8 (AMC 8) (2015)
- Scored in the 99th Percentile in the SAT, the PSAT, and the ACT (2016)

About The Authors

Jesse Doan

Jesse Doan started participating in math competitions in sixth grade. He earned perfect scores on the math section of the SAT and the PSAT. As an eighth grader, he scored a perfect 5 on the AP Calculus BC exam. Jesse is from the island of Maui, Hawaii and will be attending Stanford University in the fall of 2017.

Below are some of Jesse's achievements:

- United States of America Math Olympiad (USAMO) Qualifier (2017)
- United States of America Junior Math Olympiad (USAJMO) Qualifier (2015)
- First Place Individual in Hawaii, American Invitational Mathematical Examination (AIME) (2015, 2017)
- First Place Individual in Hawaii, American Mathematics Competitions 12 (AMC 12) (2016, 2017)
- First Place Individual in Hawaii, American Mathematics Competitions 10 (AMC 10) (2014, 2015)
- Captain of the First Place Team, Hawaii State Math Bowl (2014, 2015, 2016, 2017)
- 48th Place Individual in the USA, MATHCOUNTS Nationals (2013)
- MATHCOUNTS Nationals Alumnus (2012, 2013)
- First Place Individual in Hawaii, MATHCOUNTS State Competition (2013)
- First Place Individual in Hawaii, American Mathematics Competitions 8 (AMC 8) (2012)

For the Rising Math Olympians

The Ultimate Handbook for Winning Math Competitions in Elementary and Middle School

For the Rising Math Olympians contains over 500 examples and problems in Number Theory, Algebra, Counting & Probability, and Geometry that are frequently tested in math competitions. Each chapter contains concepts with detailed explanations, examples with step-by-step solutions, and review problems to reinforce the students' understanding.

This book is written for beginning mathletes who are interested in learning advanced problem solving and critical thinking skills in preparation for elementary and middle school math competitions.

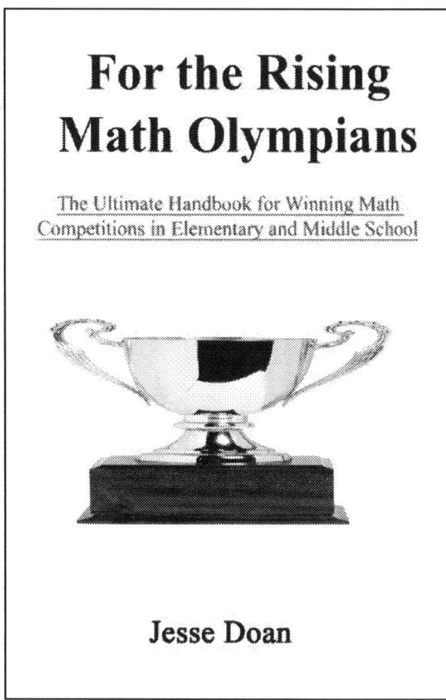

For the Rising Math Olympians

ISBN-13: 978-1536991079
ISBN-10: 1536991074

To purchase a copy of **For the Rising Math Olympians**, they are available at **Amazon.com**.

For more information, please visit our website at **risingmatholympians.weebly.com** and check out our Facebook Page at **facebook.com/RisingMathOlympians**.

Made in the USA
Columbia, SC
19 March 2025

55366098R00137